INDIAN MATHEMATICS

INDIAN MATHEMATICS

G. R. KAYE

GYAN

Published by

GYAN

Gyan Publishing House
5, Ansari Road
Daryaganj, New Delhi-110002
Phone: 011-47034999, 9811692060
E-mail: books@gyanbooks.com

Distribution Network
gyanbooks.com
India, USA, Canada, UK, Australia

ISBN: 978-93-7014-156-8 (PB)

First Published, 1915

2nd Impression 2025

Printed at: Gyan Press, Delhi.

INDIAN MATHEMATICS
Author: G. R. KAYE

INDIAN MATHEMATICS

G. R. KAYE

CALCUTTA & SIMLA
THACKER, SPINK & CO

1915

CONTENTS.

I.

1. THE orientalists who exploited Indian history and literature about a century ago were not always perfect in their methods of investigation and consequently promulgated many errors. Gradually, however, sounder methods have obtained and we are now able to see the facts in more correct perspective. In particular the early chronology has been largely revised and the revision in some instances has important bearings on the history of mathematics and allied subjects. According to orthodox Hindu tradition the *Sūrya Siddhānta*, the most important Indian astronomical work, was composed over two million years ago! Bailly, towards the end of the eighteenth century, considered that Indian astronomy had been founded on accurate observations made thousands of years before the Christian era. LAPLACE, basing his arguments on figures given by BAILLY considered that some 3,000 years B.C. the Indian astronomers had recorded actual observations of the planets *correct to one second*; PLAYFAIR eloquently supported BAILLY's views; Sir WILLIAM JONES argued that correct observations must have been made at least as early as 1181 B.C.; and so on; but with the researches of COLEBROOKE, WHITNEY, WEBER, THIBAUT, and others more correct views were introduced and it was proved that the records used by Bailly were quite modern and that the actual period of the composition of the original *Sūrya Siddhānta* was not earliar than A.D. 400.

It may, indeed, be generally stated that the tendency of the early orientalists was towards antedating and this tendency is exhibited in discussions connected with two notable works, the *Śulvasūtras* and the *Bakhshālī* arithmetic, the dates of which are not even yet definitely fixed.

2. In the 16th century, A.D., Hindu tradition ascribed the invention of the nine figures with the device of place to make them suffice for all numbers 'to the beneficent Creator of the universe'; and this was accepted as evidence of the very great antiquity of the system! This is a particular illustration of an attitude that was quite general, for early Indian works claim either to be directly revealed or of divine origin. One consequence of this attitude is that we find absolutely no references to foreign origins or foreign influence.* We have, however, a great deal of direct evidence that proves conclusively that foreign influence was very real indeed—Greek and Roman coins, coins with Greek and Indian inscriptions, Greek technical terms, etc., etc.; and the implication of considerable foreign influence occurs in certain classes of literature and also in the archæological remains of the north-west of India. One of the few references to foreigners is given by Vahrāha Mihira who acknowledged that the Greeks knew something of astrology; but although he gives accounts of the *Romaka* and the *Pauliśa siddhāntas* he never makes any direct acknowledgment of western influence.

* It may be noted that beyond the vague pseudo-prophetic references in the *Purāṇas*, no early Indian writer mentions the invasion of Alexander the Great.

II.

3. For the purpose of discussion three periods in the history of Hindu mathematics may be considered :

(I) The *S'ulvasūtra* period with upper limit *c.* A.D. 200 ;

(II) The astronomical period *c.* A.D. 400—600.

(III) The Hindu mathematical period proper, A.D. 600—1200.

Such a division into periods does not, of course, perfectly represent the facts, but it is a useful division and serves the purposes of exposition with sufficient accuracy. We might have prefixed an earlier, or Vedic, period but the literature of the Vedic age does not exhibit anything of a mathematical nature beyond a few measures and numbers used quite informally. It is a remarkable fact that the second and third of our periods have no connection whatever with the first or *S'ulvasūtra* period. The later Indian mathematicians completely ignored the mathematical contents of the *S'ulvasūtras.* They not only never refer to them but do not even utilise the results given therein. We can go even further and state that no Indian writer earlier than the nineteenth century is known to have referred to the *S'ulvasūtras* as containing anything of mathematical value. This disconnection will be illustrated as we proceed and it will be seen that the works of the third period may be considered as a direct development from those of the second.

4. THE S'ULVASUTRA PERIOD.—The term *S'ulvasūtra* means ' the rules of the cord ' and is the name given to the supplements of the *Kalpasūtras* which treat of the construction of sacrificial altars. The period in which the *S'ulvasūtras* were composed has been variously fixed by various

authors. MAX MÜLLER gave the period as lying between 500 and 200 B.C.; R. C. DUTT gave 800 B.C.; BÜHLER places the origin of the Apastamba school as probably somewhere within the last four centuries before the Christian era, and Baudhāyana somewhat earlier; MACDONNELL gives the limits as 500 B.C. and A.D. 200, and so on. As a matter of fact the dates are not known and those suggested by different authorities must be used with the greatest circumspection. It must also be borne in mind that the contents of the *S'ulvasūtras*, as known to us, are taken from quite modern manuscripts; and that in matters of detail they have probably been extensively edited. The editions of Āpastamba, Baudhāyana and Kātyāyana which have been used for the following notes, indeed, differ from each other to a very considerable extent.

The *S'ulvasūtras* are not primarily mathematical but are rules ancillary to religious ritual—they have not a mathematical but a religious aim. No proofs or demonstrations are given and indeed in the presentation there is nothing mathematical beyond the bare facts. Those of the rules that contain mathematical notions relate to (1) the construction of squares and rectangles, (2) the relation of the diagonal to the sides, (3) equivalent rectangles and squares, (4) equivalent circles and squares.

5. In connection with (1) and (2) the PYTHAGOREAN theorem is stated quite generally. It is illustrated by a number of examples which may be summarised thus:

ĀPASTAMBA.	BAUDHĀYANA.
$3^2+4^2=5^2$	$3^2+4^2=5^2$
$12^2+16^2=20^2$	$5^2+12^2=13^2$
$15^2+20^2=25^2$	$8^2+15^2=17^2$
$5^2+12^2=13^2$	$7^2+24^2=25^2$
$15^2+36^2=39^2$	$12^2+35^2=37^2$
$8^2+15^2=17^2$	$15^2+36^2=39^2$
$12^2+35^2=37^2$	

Kātyāyana gives no such rational examples but gives (with Āpastamba and Baudhāyana) the hypotenuse corresponding to sides equal to the side and diagonal of a square, *i.e.*, the triangle a, $a\sqrt{2}$, $a\sqrt{3}$, and he alone gives $1^2+3^2=10$, and $2^2+6^2=40$. There is no indication that the *Śulvasūtra* rational examples were obtained from any general rule. Incidentally is given an arithmetical value of the diagonal of a square which may be represented by

$$\sqrt{2} = 1 + \tfrac{1}{3} + \frac{1}{3.4} - \frac{1}{3.4.34}.$$

This has been much commented upon but, given a scale of measures based upon the change ratios 3, 4, and 34 (and Baudhāyana actually gives such a scale) the result is only an expression of a direct measurement ; and for a side of six feet it is accurate to about $\frac{1}{7}$th of an inch ; or it is possible that the result was obtained by the approximation $\sqrt{a^2+b} = a + \dfrac{b}{2a}$ by TANNERY'S R- process, but it is quite certain that no such process was known to the authors of the *Śulvasūtras.* The only noteworthy character of the fraction is the form with its unit numerators. Neither the value itself nor this form of fraction occurs in any later Indian work.

There is one other point connected with the PYTHAGOREAN theorem to be noted, *viz.*, the occurrence of an indication of the formation of a square by the successive addition of gnomons. The text relating to this is as follows :

> "Two hundred and twenty-five of these bricks constitute the sevenfold *agni* with *aratni* and *pradesa.*"

> "To these sixty-four more are to be added. With these bricks a square is formed. The side of the square consists of sixteen bricks. Thirty-three

> bricks still remain and these are placed on all sides round the borders."

This subject is never again referred to in Indian mathematical works.

The questions (*a*) whether the Indians of this period had completely realised the generality of the PYTHAGOREAN theorem, and (*b*) whether they had a sound notion of the irrational have been much discussed; but the ritualists who composed the *S'ulvasūtras* were not interested in the PYTHAGOREAN theorem beyond their own actual wants, and it is quite certain that even as late as the 12th century no Indian mathematician gives evidence of a complete understanding of the irrational. Further, at no period did the Indians develop any real theory of geometry, and a comparatively modern Indian work denies the possibility of any proof of the PYTHAGOREAN theorem other than experience.

The fanciful suggestion of BÜRK that possibly PYTHAGORAS obtained his geometrical knowledge from India is not supported by any actual evidence. The Chinese had acquaintance with the theorem over a thousand years B.C., and the Egyptians as early as 2000 B.C.

6. Problems relating to equivalent squares and rectangles are involved in the prescribed altar constructions and consequently the *S'ulvasūtras* give constructions, by help of the PYTHAGOREAN theorem, of

(1) a square equal to the sum of two squares;
(2) a square equal to the difference of two squares;
(3) a rectangle equal to a given square;
(4) a square equal to a given rectangle;
(5) the decrease of a square into a smaller square.

Again we have to remark the significant fact that none of these geometrical constructions occur in any later Indian work. The first two are direct geometrical applications of the rule $c^2=a^2+b^2$; the third gives in a geometrical form the sides of the rectangle as $a\sqrt{2}$ and $\frac{a\sqrt{2}}{2}$; the fourth rule gives a geometrical construction for $ab = \left(b+\frac{a-b}{2}\right)^2 - \left(\frac{a-b}{2}\right)^2$ and corresponds to EUCLID, II, 5; the fifth is not perfectly clear but evidently corresponds to EUCLID, II, 4.

7. THE CIRCLE.—According to the altar building ritual of the period it was, under certain circumstances, necessary to square the circle, and consequently we have recorded in the *Śulvasūtras* attempts at the solution of this problem, and its connection with altar ritual reminds us of the celebrated DELIAN problem. The solutions offered are very crude although in one case there is pretence of accuracy. Denoting by a the side of the square and by d the diameter of the circle whose area is supposed to be a^2 the rules given may be expressed by

$$(\alpha)\quad d=a+\tfrac{1}{3}(a\sqrt{2}-a)$$

$$(\beta)\quad a=d-\tfrac{2}{15}d$$

$$(\gamma)\quad a=d\left(1-\frac{1}{8}+\frac{1}{8.29}-\frac{1}{8.29.6}+\frac{1}{8.29.6.8}\right)$$

Neither of the first two rules, which are given by both Āpastamba and Baudhāyana, is of particular value or interest. The third is given by Baudhāyana *only* and is evidently obtained from (α) by utilising the value for $\sqrt{2}$ given in paragraph 5 above. We thus have

$$\frac{a}{d}=\frac{3}{2+\sqrt{2}}=\frac{3}{2+\frac{577}{408}}=\frac{1224}{1393}$$

$$=1-\frac{1}{8}+\frac{1}{8.29}-\frac{1}{8.29.6}+\frac{1}{8.29.6.8}-\frac{41}{8.29.6.8.1393}$$

which, neglecting the last term, is the value given in rule (γ). This implies a knowledge of the process of converting a fraction into partial fractions with unit numerators, a knowledge most certainly not possessed by the composers of the *Śulvasūtras*; for as THIBAUT says there is nothing in these rules which would justify the assumption that they were expert in long calculations; and there is no indication in any other work that the Indians were ever acquainted with the process and in no later works are fractions expressed in this manner.

It is worthy of note that later Indian mathematicians record no attempts at the solution of the problem of squaring the circle and never refer to those recorded in the *S'ulvasūtras*.

III.

A.D. 400 TO 600.

8. There appears to be no connecting link between the *S'ulvasūtra* mathematics and later Indian developments of the subject. Subsequent to the *S'ulvasūtras* nothing further is recorded until the introduction into India of western astronomical ideas.* In the sixth century A.D. VARĀHA MIHIRA wrote his *Pañcha Siddhāntikā* which gives a summary account of the five most important astronomical works then in use. Of these the *Sūrya Siddhānta*, which was probably composed in its original form not earlier than A.D. 400, afterwards became the standard work. VARĀHA MIHIRA's collection is the earliest and most authentic account we have of what may be termed the scientific treatment of astronomy in India. "Although," writes THIBAUT, "not directly stating that the Hindus learned from the Greeks, he at any rate mentions certain facts and points of doctrine which suggest the dependence of Indian astronomy on the science of Alexandria; and, as we know already from his astrological writings, he freely employs terms of undoubted Greek origin."

VARĀHA MIHIRA writes:—"There are the following Siddhāntas—the Pauliśa, the Romaka, the Vasishtha, the Saura and the Paitamaha....... The Siddhānta made by Pauliśa is accurate, near to it stands the Siddhānta proclaimed by Romaka, more accurate is the Sāvitra (Sūrya). The two remaining ones are far from the truth."

---

* This has a somewhat important bearing on the date of the *S'ulvasūtras*. If, for example, the date of their composition were accepted as 500 B.C. a period of nearly 1,000 years, absolutely blank as far as mathematical notions are concerned, would have to be accounted for.

9. The *Pañcha Siddhāntikā* contains material of considerable mathematical interest and from the historical point of view of a value not surpassed by that of any later Indian works. The mathematical section of the *Pauliśa Siddhānta* is perhaps of the most interest and may be considered to contain the essence of Indian trigonometry. It is as follows :—

"(1) The square-root of the tenth part of the square of the circumference, which comprises 360 parts, is the diameter. Having assumed the four parts of a circle the sine of the eighth part of a sign [is to be found].

"(2) Take the square of the radius and call it the constant. The fourth part of it is [the square of] Aries. The constant square is to be lessened by the square of Aries. The square-roots of the two quantities are the sines.

"(3) In order to find the rest take the double of the arc, deduct it from the quarter, diminish the radius by the sine of the remainder and add to the square of half of that the square of half the sine of double the arc. The square-root of the sum is the desired sine."

[The eighth part of a "sign" (=30°) is 3° 45′ and by "Aries" is indicated the first "sign" of 30°.]

The rules given may be expressed in our notation (for unit radius) as

$$(1)\quad \pi=\sqrt{10}\qquad (2)\quad \text{Sin } 30^\circ=\tfrac{1}{2},\quad \text{Sin } 60^\circ=\sqrt{1-\tfrac{1}{4}},$$

$$(3)\quad \text{Sin}^2\gamma=\left(\frac{\text{Sin } 2\gamma}{2}\right)^2+\left(\frac{1-\text{Sin }(90-2\gamma)}{2}\right)^2$$

They are followed by a table of 24 sines progressing by intervals of 3° 45′ obviously taken from PTOLEMY's table of chords. Instead, however, of dividing the radius into 60 parts, as did Ptolemy, Pauliśa divides it into 120 parts; for as $\sin \frac{a}{2} = \frac{\text{chord } a}{2}$ this division of the radius enabled him to convert the table of chords into sines without numerical change. Āryabhata gives another measure for the radius (3438′) which enabled the sines to be expressed in a sort of circular measure.

We thus have three distinct stages :

(*a*) The chords of PTOLEMY, or ch'da, with $r=60$

(*b*) The Pauliśa sine or $\sin \frac{a}{2} = \frac{\text{ch'd } a}{2}$, with $r=120$

(*c*) The Āryabhata sine or $\sin \frac{a}{2} = \frac{3}{\pi} \cdot \frac{\text{ch'd } a}{2}$

with $r=3438'$

To obtain (*c*) the value of π actually used was $\frac{600}{191}$ ($=3.14136$)

Thus the earliest known record of the use of a sine function occurs in the Indian astronomical works of this period. At one time the invention of this function was attributed to el-Battânî [*A.D.* 877—919] and although we now know this to be incorrect we must acknowledge that the Arabs utilised the invention to a much more scientific end than did the Indians.

In some of the Indian works of this period an interpolation formula for the construction of the table of sines is given. It may be represented by

$$\Delta_{n+1} = \Delta_n - \frac{\text{Sin } n.\, a}{\text{Sin } a} \text{ where } \Delta_n = \text{Sin } n.\, a - \text{Sin } (n-1)a,$$

This is given ostensibly for the formation of the table, but the table actually given cannot be obtained from the formula.

10. ĀRYABHATA.—Tradition places Āryabhaṭa (born A.D. 476) at the head of the Indian mathematicians and indeed he was the first to write formally on the subject.*

* Although Āryabhata's *Ganita*, as first published by Kern, is generally accepted as authentic, there is an element of doubt in the matter.

He was renowned as an astronomer and as such tried to introduce sounder views of that science but was bitterly opposed by the orthodox. The mathematical work attributed to him consists of thirty-three couplets into which is condensed a good deal of matter. Starting with the orders of numerals he proceeds to evolution and involution, and areas and volumes. Next comes a semi-astronomical section in which he deals with the circle, shadow problems, etc. ; then a set of propositions on progressions followed by some simple algebraic identities. The remaining rules may be termed practical applications with the exception of the very last which relates to indeterminate equations of the first degree. Neither demonstrations nor examples are given, the whole text consisting of sixty-six lines of bare rules so condensed that it is often difficult to interpret their meaning. As a mathematical treatise it is of interest chiefly because it is some record of the state of knowledge at a critical period in the intellectual history of the civilised world ; because, as far as we know, it is the earliest Hindu work on pure mathematics ; and because it forms a sort of introduction to the school of Indian mathematicians that flourished in succeeding centuries.

Āryabhaṭa's work contains one of the earliest records known to us of an attempt at a general solution of indeterminates of the first degree by the continued fraction process. The rule, as given in the text, is hardly coherent but there is no doubt as to its general aim. It may be considered as forming an introduction to the somewhat marvellous development of this branch of mathematics that we find recorded in the works of Brahmagupta and Bhāskara. Another noteworthy rule given by Āryabhaṭa is the one which contains an extremely accurate value of the ratio of the circumference of a circle to the diameter, *viz.*, $\pi = 3 \frac{177}{1250}$ ($=3{\cdot}1416$) ; but it is rather extraordinary that Āryabhaṭa himself never utilised this value, that it was not used by any other Indian mathematician before the 12th century and that no Indian writer quotes Āryabhaṭa as recording this value. Other noteworthy points are the rules relating to volumes of solids which contain some remarkable inaccuracies, *e.g.*, the volume of a pyramid is given as *half* the product of the height and the base ; the volume of a sphere is stated to be the product of the area of a circle (of the same radius as the sphere) and the root of this area, or $\pi^{\frac{3}{2}}\gamma^3$. Similar errors were not uncommon in later Indian works. The rule known as the *epanthem* occurs in Āryabhaṭa's work and there is a type of definition that occurs in no other Indian work, *e.g.*, " The product of three equal numbers is a cube and it also has twelve edges."

IV.

A.D. 600—1200.

11. Aryabhata appears to have given a definite bias to Indian mathematics, for following him we have a series of works dealing with the same topics. Of the writers themselves we know very little indeed beyond the mere names but some if not all the works of the following authors have been preserved :

Brahmagupta	..	.. born A.D. 598.
Mahāvīra	..	.. ? 9th century.
Śrīdhara	..	.. born A.D. 991.
Bhāskara	..	.. born A.D. 1114.

Bhāskara is the most renowned of this school, probably undeservedly so, for Brahmagupta's work is possibly sounder mathematically and is of much more importance historically. Generally these writers treat of the same topics—with a difference—and Brahmagupta's work appears to have been used by all the others. Bhāskara mentions another mathematician, Padmanābha, but omits from his list Mahāvīra.

One of the chief points of difference is in the treatment of geometry. Brahmagupta deals fairly completely with cyclic quadrilaterals while the later writers gradually drop this subject until by the time of Bhāskara it has ceased to be understood.

The most interesting characteristics of the works of this period are the treatment of :

(*i*) indeterminate equations ;

(*ii*) the rational right-angled triangle ;

and (*iii*) the perfunctory treatment of pure geometry.

Of these topics it will be noted that the second was dealt with to some extent in the *S'ulvasūtras* ; but a close examination seems to show that there is no real connection and that the writers of the third period were actually ignorant of the results achieved by Baudhāyana and Āpastamba.

12. **Indeterminate Equations.** The interesting names and dates connected with the early history of indeterminates in India are :

				cir. A.D.
Diophantus	..	..	„	360
Hypatia	..	..		400
Āryabhaṭa	..	..	born	476
Brahmagupta	..	..	„	598
Bhāskara	..	..	„	1114

That we cannot fill up the gap between Diophantus and Āryabhaṭa with more than the mere name of Hypatia is probably due to the fanatic ignorance and cruelty of the early Alexandrian Christians rather than the supposed destruction of the Alexandrian library by the Muhammadans. It would be pleasant to conceive that in the Indian works we have some record of the advances made by Hypatia, or of the contents of the lost books of Diophantus—but we are not justified in indulging in more than the mere fancy. The period is one of particular interest. The murder of Hypatia (A.D. 415), the imprisonment and execution of Boethius (A.D. 524), the closing of the Athenean schools in A.D. 530 and the fall of Alexandria in 640 are events full of suggestions to the historian of mathematics. It was during this period also that Damascius, Simplicius (mathematicians of some repute) and others of the schools of Athens, having heard that Plato's ideal form of government was actually realised under Chosroes I in Persia, emigrated thither (c. A.D

532). They were naturally disappointed but the effect of their visit may have been far greater than historical records show.

13. The state of knowledge regarding indeterminate equations in the west at this period is not definitely known. Some of the works of Diophantus and all those of Hypatia are lost to us ; but the extant records show that the Greeks had explored the field of this analysis so far as to achieve rational solutions (not necessarily integral) of equations of the first and second degree and certain cases of the third degree. The Indian works record distinct advances on what is left of the Greek analysis. For example they give rational *integral* solutions of

$$(A)\quad ax \pm by = c$$

$$(B)\quad Du^2 + 1 = t^2$$

The solution of (A) is only roughly indicated by Ārya-bhaṭa but Brahmagupta's solution (for the positive sign) is practically the same as

$$x = \pm cq - bt, \qquad y = \mp cp + at$$

where t is zero or any integer and p/q is the penultimate convergent of a/b.

The Indian methods for the solution of

$$Du^2 + 1 = t^2$$

may be summarised as follows :

If $Da^2 + b = c^2$ and $D\alpha^2 + \beta = \gamma^2$ then will

$$(a)\quad D(c\alpha \mp \gamma a)^2 + b\beta = (c\gamma \pm a\alpha D)^2$$

$$(b)\quad D\left(\frac{c+ar}{b}\right)^2 \pm \frac{r^2 - D}{b} = \left(\frac{Da+cr}{b}\right)^2$$

where r is any suitable integer.

Also

$$(c)\quad D\left(\frac{2n}{D-n^2}\right)^2 + 1 = \left(\frac{D+n^2}{D-n^2}\right)^2$$

where n is any assumed number.

The complete integral solution is given by a combination (a) and (b) of which the former only is given by Brahmagupta, while both are given by Bhāskara (five centuries later). The latter designates (*a*) the 'method by composition' and (*b*) the 'cyclic method.' These solutions are alone sufficient to give to the Indian works an important place in the history of mathematics. Of the 'cyclic method' (*i.e.*, the combination of (*a*) and (*b*)) HANKEL says, "It is beyond all praise: it is certainly the finest thing achieved in the theory of numbers before Lagrange." He attributed its invention to the Indian mathematicians, but the opinions of the best modern authorities (*e.g.*, TANNERY, CANTOR, HEATH) are rather in favour of the hypothesis of ultimate Greek origin.

The following conspectus of the indeterminate problems dealt with by the Indians will give some idea of their work in this direction; and although few of the cases actually occur in Greek works now known to us the conspectus significantly illustrates a general similarity of treatment.

*(1) $ax \pm by = c$

(2) $ax + by + cz + \dots = l$

(3) $x \equiv a_1 \; Mod. \; b_1 \equiv \dots \equiv a_4 \; Mod. \; b_4$

(4) $Ax + By + Cxy = D$

(5) $Du^2 + 1 = t^2$

(6) $Du^2 - 1 = t^2$

(7) $Du^2 \pm s = t^2$

(8) $D^2u^2 \pm s = t^2$

(9) $u^2 + s = at^2$

(10) $Du^2 \pm au = t^2$

(11) $s - Du^2 = t^2$

(12) $Du + s = t^2$

(13) $x \pm a = s^2, \quad x \pm b = t^2$

* Of these only numbers 1—5, 7, 8, 12—14 occur before the twelfth century.

(14) $ax + 1 = s^2, \quad bx + a = t^2$

(15) $2(x^2 - y^2) + 3 = s^2, \quad 3(x^2 - y^2) + 3 = t^2$

(16) $ax^2 + by^2 = s^2, \quad ax^2 - by^2 + 1 = t^2$

(17) $x^2 + y^2 \pm 1 = s^2, \quad x^2 - y^2 \pm 1 = t^2$

(18) $x^2 - a \equiv x^2 - b \equiv o \; Mod. \; c$

(19) $ax^2 + b \equiv o \; Mod. \; c$

(20) $x + y = s^2, \quad x - y = t^2, \quad xy = u^3$

(21) $x^3 + y^3 = s^2, \quad x^2 + y^2 = t^3$

(22) $x - y = s^3, \quad x^2 + y^2 = t^3$

(23) $x + y = s^2, \quad x^3 + y^2 = t^2$

(24) $x^3 + y^2 + xy = s^2, \quad (x + y)s + 1 = t^2$

(25) $ax + 1 = s^3, \quad as^2 + 1 = t^2$

(26) $wxyz = a(w + x + y + z)$

(27) $x^3 - a \equiv o \; Mod. \; b$

(28) $x + y + 3 = s^2, x - y + 3 = t^2, x^2 + y^2 - 4 = u^2,$

$x^2 - y^2 + 12 = v^2, \quad \frac{xy}{2} + y = w^3,$

$s + t + u + v + w + 2 = z^2$

(29) $\sqrt[3]{\frac{xy+y}{2}} + \sqrt{x^2+y^2} + \sqrt{x+y+2} + \sqrt{x-y+2} +$

$\sqrt{x^2-y^2+8} = t^2$

(30) $w + 2 = a^2, \; x + 2 = b^2, \; y + 2 = c^2, \; z + 2 = d^2,$

$wx + 18 = e^2, \quad xy + 18 = f^2, \quad yz + 18 = g^2,$

$a + b + c + d + e + f + g + 11 = 13.$

14. Rational right-angled triangles.—The Indian mathematicians of this period seem to have been particularly attracted by the problem of the rational right-angled triangle and give a number of rules for obtaining integral solutions. The following summary of the various rules relating to this problem shows the position of the Indians fairly well.

	A	B	$\sqrt{A^2+B^2}$	Authorities.
i	n	$\frac{n^2-1}{2}$	$\frac{n^2+1}{2}$	Pythagoras (according to Proclus)
ii	$\sqrt{m\,n}$	$\frac{m\ n}{2}$	$\frac{m+n}{2}$	Plato (according to Proclus)
iii	$2\,mn$	m^2-n^2	m^2+n^2	Euclid and Diophantus.
iv	$\frac{m\,(n^2-1)}{n^2+1}$	$\frac{2mn}{n^2+1}$	m	Diophantus.
v	$2\,mn$	m^2-n^2	m^2+n^2	Brahmagupta and Mahāvīra.
vi	$\sqrt{m}$	$\frac{1}{2}\left(\frac{m}{n}-n\right)$	$\frac{1}{2}\left(\frac{m}{n}+n\right)$	Brahmagupta, Mahāvīra, and Bhāskara.
vii	m	$\frac{2\,mn}{n^2-1}$	$\frac{m\,(n^2+1)}{n^2-1}$	Bhāskara.
viii	$\frac{m\,(n^2-1)}{n^2+1}$	$\frac{2\,mn}{n^2+1}$	m	Bhāskara.
ix	$2\,l\,mn$	$l\,(m^2-n^2)$	$l\,(m^2+n^2)$	General formula.

Mahāvīra gives many examples in which he employs formula (V) of which he terms m and n the 'elements.' From given elements he constructs triangles and from given triangles he finds the elements, *e.g.*, "What are the 'elements' of the right-angled triangle (48, 55, 73)? Answer: 3, 8."

Other problems connected with the rational right-angled triangle given by Bhāskara are of some historical interest: *e.g.*, (1) The sum of the sides is 40 and the area 60, (2) The sum of the sides is 56 and their product 7×600, (3) The area is numerically equal to the hypotenuse, (4) The area is numerically equal to the product of the sides.

15. The geometry of this period is characterised by:

(1) Lack of definitions, etc.;

(2) Angles are not dealt with at all;

(3) There is no mention of parallels and no theory of proportion;

(4) Traditional inaccuracies are not uncommon;

(5) A gradual decline in geometrical knowledge is noticeable.

On the other hand, we have the following noteworthy rules relating to cyclic quadrilaterals—

$$(i)\quad Q=\sqrt{(s-a)(s-b)(s-c)(s-d)}$$

$$(ii)\quad x^2=(ad+bc)(ac+bd)/(ab+cd)$$

$$y^2=(ab+cd)(ac+bd)/(ad+bc)$$

where x and y are the diagonals of the cyclic quadrilateral (a, b, c, d). This (*ii*) is sometimes designated as 'Brahmagupta's theorem'.

16. The absence of definitions and indifference to logical order sufficiently differentiate the Indian geometry from that of the early Greeks; but the absence of what may be termed a theory of geometry hardly accounts for the complete absence of any reference to parallels and angles. Whereas on the one hand the Indians have been credited with the invention of the sine function, on the other there is no evidence to show that they were acquainted with even the most elementary theorems (as such) relating to angles.

The presence of a number of incorrect rules side by side with correct ones is significant. The one relating to the area of triangles and quadrilaterals, *viz.*, the area is equal to the product of half the sums of pairs of opposite sides, strangely enough occurs in a Chinese work of the 6th century as well as in the works of Ahmes, Brahmagupta, Mahāvīra, Boethius and Bede. By Mahāvīra, the idea on which it is based—that the area is a function of the perimeter—is further emphasized. Āryabhaṭa gives an incorrect rule for the volume of a pyramid ; incorrect rules for the volume of a sphere are common to Āryabhaṭa, S'rīdhara and Mahāvīra. For cones all the rules assume that $\pi=3$. Mahāvīra gives incorrect rules for the circumference and area of an ellipse and so on.

17. Brahmagupta gives a fairly complete set of rules dealing with the cyclic quadrilateral and either he or the mathematician from whom he obtained his material had a definite end in view—the construction of a cyclic quadrilateral with rational elements.—The commentators did not fully appreciate the theorems, some of which are given in the works of Mahāvīra and S'rīdhara ; and by the time of Bhāskara they had ceased to be understood. Bhāskara indeed condemns them outright as unsound. "How can a person" he says "neither specifying one of the perpendiculars, nor either of the diagonals, ask the rest ? Such a questioner is a blundering devil and still more so is he who answers the question."

Besides the two rules (*i*) and (*ii*) already given in paragraph 15, Brahmagupta gives rules corresponding to the formula

(*iii*) $2r=\frac{a}{\sin A}$ etc., and

(*iv*) If $a^2+b^2=c^2$ and $\alpha^2+\beta^2=\gamma^2$ then the quadrilateral $(a\gamma,\ c\beta,\ b\gamma,\ c\alpha)$ is cyclic and has its diagonals at right angles.

This figure is sometimes termed "Brahmagupta's trapezium." From the triangles (3, 4, 5) and (5, 12, 13) a commentator obtains the quadrilateral (39, 60, 52, 25), with diagonals 63 and 56, etc. He also introduces a proof of Ptolemy's theorem and in doing this follows Diophantus (*iii*, 19) in constructing from triangles (a, b, c) and $(\alpha, \beta, \gamma,)$ new triangles $(a\gamma, b\gamma, c\gamma,)$ and $(\alpha c, \beta c, \gamma c,)$ and uses the actual examples given by Diophantus, namely (39, 52, 65) and (25, 60, 65).

18. An examination of the Greek mathematics of the period immediately anterior to the Indian period with which we are now dealing shows that geometrical knowledge was in a state of decay. After Pappus (c. A.D. 300) no geometrical work of much value was done. His successors were, apparently, not interested in the great achievements of the earlier Greeks and it is certain that they were often not even acquainted with many of their works. The high standard of the earlier treatises had ceased to attract, errors crept in, the style of exposition deteriorated and practical purposes predominated. The geometrical work of Brahmagupta is almost what one might expect to find in the period of decay in Alexandria. It contains one or two gems but it is not a scientific exposition of the subject and the material is obviously taken from western works.

V.

19. We have, in the above notes, given in outline the historically important matters relating to Indian mathematics. For points of detail the works mentioned in the annexed bibliography should be consulted ; but we here briefly indicate the other contents of the Indian works, and in the following sections we shall refer to certain topics that have achieved a somewhat fictitious importance, to the personalities of the Indian mathematicians and to the relations between the mathematics of the Chinese, the Arabs and the Indians.

20. Besides the subjects already mentioned BRAHMAGUPTA deals very briefly with the ordinary arithmetical operations, square and cube-root, rule of three, etc. ; interest, mixtures of metals, arithmetical progressions, sums of the squares of natural numbers ; geometry as already described but also including elementary notions of the circle ; elementary mensuration of solids, shadow problems, negative and positive qualities, cipher, surds, simple algebraic identities ; indeterminate equations of the first and second degree, which occupy the greater portion of the work, and simple equations of the first and second degrees which receive comparatively but little attention.

MAHĀVĪRA'S work is fuller but more elementary on the whole. The ordinary operations are treated with more completeness and geometrical progressions are introduced ; many problems on indeterminates are given but no mention is made of the ' cyclic method ' and it contains no formal algebra. It is the only Indian work that deals with ellipses (inaccurately).

The only extant work by S'RĪDHARA is like Mahāvīra's but shorter ; but he is quoted as having dealt with quadratic equations, etc.

Bhāskara's *Līlāvatī* is based on S'rīdhara's work and, besides the topics already mentioned, deals with combinations, while his *Vīja-gaṇita*, being a more systematic exposition of the algebraical topics dealt with by Brahmagupta, is the most complete of the Indian algebras.

After the time of Bhāskara (born A.D. 1114) no Indian mathematical work of historical value or interest is known. Even before his time deterioration had set in and although a "college" was founded to perpetuate the teaching of Bhāskara it, apparently, took an astrological bias.

21. The Indian method of stating examples—particularly those involving algebraic equations—are of sufficient interest to be recorded here. The early works were rhetorical and not symbolical at all and even in modern times the nearest approach to a symbolic algebra consists of abbreviations of special terms. The only real symbol employed is the negative sign of operation, which is usually a dot placed above or at the side of the quantity affected. In the Bakhshāli Ms., a cross is used in place of the dot as the latter in the S'ārada script is employed to indicate cipher or nought.

The first mention of special terms to represent unknown quantities occurs in Bhāskara's *Vīja-gaṇita* which was written in the twelfth century of our era. Bhāskara says : "As many as (*yāvat tāvat*) and the colours 'black (*kālaka*), blue (*nīlaka*), yellow (*pītaka*) and red (*lohitaka*)' and others besides these have been selected by ancient teachers* for names of values of unknown quantities."

The term *yāvat tāvat* is understandable and so is the use of colours but the conjunction is not easy to understand. The use of two such diverse types as *yāvat tāvat* and *kālaka*

*Not Indians.

(generally abbreviated to *yā* and *kā*) in one system suggests the possibility of a mixed origin. It is possible that the former is connected with Diophantus' definition of the unknown quantity, *plēthos monádon aoriston*, *i.e.*, 'an undefined (or unlimited) number of units.' To pass from 'an unlimited number' to 'as many as' requires little imagination. Diophantus had only one symbol for the unknown and if the use of *yāvat tāvat* were of Diophantine origin the Indians would have had to look elsewhere for terms for the other unknowns. With reference to the origin of the use of colours for this purpose we may point out that the very early Chinese used calculating pieces of two colours to represent positive and negative numbers.

As neither the Greeks nor the Indians used any sign for addition they had to introduce some expression to distinguish the absolute term from the variable terms. The Greeks used M° an abbreviation for *monádes* or 'units' while the Indians used *rū* for *rūpa*, a unit.

The commoner abbreviations used by the Indians are as follows :—

yā for *yāvat tāvat*, the first unknown.
kā ,, *kālaka*, the second unknown.
rū ,, *rūpa*, the absolute quantity.
va ,, *varga*, a square.
gha ,, *ghana*, a cube.
ka ,, *karaṇa*, a surd.

It is hardly appropriate to discuss Sanskrit mathematical terminology in detail here but it will not be out of place to mention a few other terms. To denote the fourth power *varga varga* is used but it occurs only once within our period. In more modern times *varga ghana ghāta*† denoted the fifth power, *varga ghana*, the sixth and so on.

† *Ghāta*=the product.

Certain Greek terms are used, e.g., *jāmitra* (*Gk.* diámetron), *kendra* (*Gk.* kentron), *trikoṇa* (*Gk.* trigōnon), *lipta* (*Gk.* leptē), *harija* (*Gk.* 'orízōn), *dramma* (*Gk.* drachmē), *dīnāra* (*Gk.* dēnārion), etc. Many of these terms, however, are borrowed from Indian astrological works which contain a considerable number of Greek terms such as *Hṛidroga* (*Gk.* (udrochoos) *Pārthona* (*Gk.* Parthénos), *āpoklima* (*Gk.* apóklima), etc., etc.

The curious may compare *pārśva* ' a rib, ' ' side ' with the Greek *pleura* ; *koṭi* which primarily means a claw or horn but is used for the perpendicular side of a triangle, with *káthetos* ; *jātya* which means ' legitimate,' ' genuine, ' but is used to denote a right-angled triangle with *orthogōnia* ; and so on.

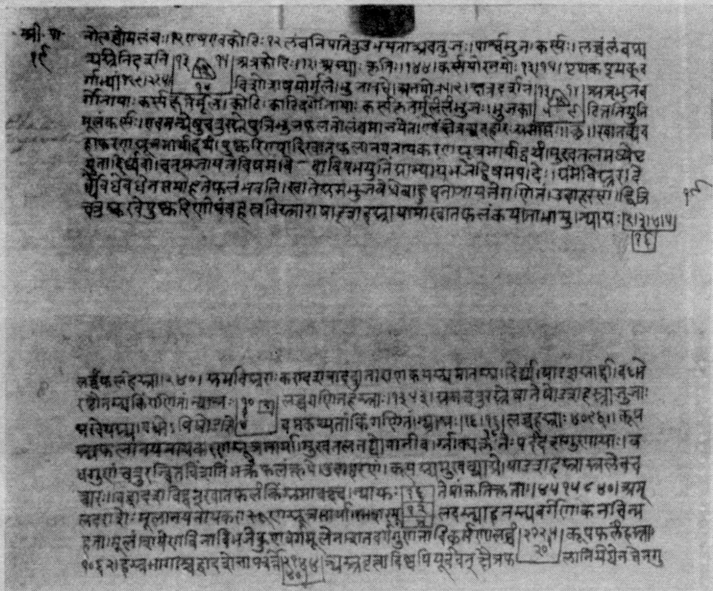

Śrīdhara's Triśatikā.

(From the copy used by Colebrooke *India Office Catalogue* 520e.)

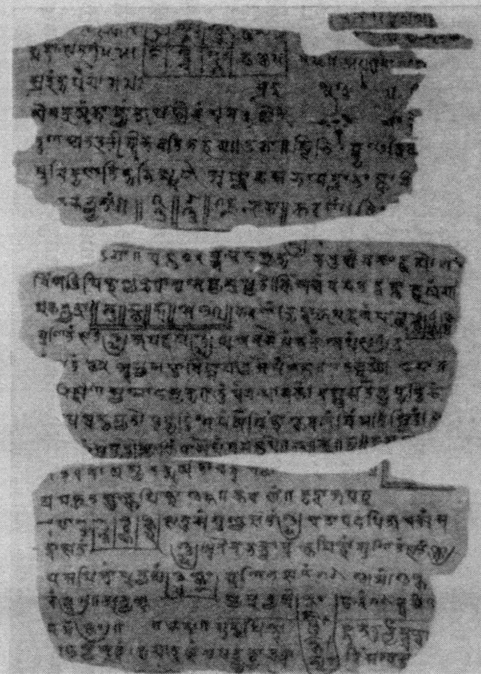

Bakhshālī Ms.

We conclude this section with a few illustrations transliterated from Sanskrit manuscripts.

<table>
<tr><th></th><th>Indian Forms.</th><th>Equivalents.</th><th>References.</th></tr>
<tr><td>1.</td><td>yā 6 rū 300
yā 10 rū $\dot{1}00$</td><td>$6x+300=10x-100$</td><td>V. 104.</td></tr>
<tr><td>2.</td><td>yāva 18 yā 0 rū 0
yāva 16 yā 9 rū 18</td><td>$18x^2+0.x+0=16x^2+9x+18$</td><td>Y. 133.</td></tr>
<tr><td>3.</td><td>yā va va 1 yā va $\dot{2}$ yā 400 rū 0
yā va va 0 yā va 0 yā 0 rū 9999</td><td>$x^4-2x^2+400x+0=0.x^4+0.x^2+0.x+9999$</td><td>V. 138.</td></tr>
<tr><td>4.</td><td>yā 197 kā $\dot{1}644$ nī $\dot{1}$ rū 0
yā 0 kā 0 nī 0 rū 6302</td><td>$197x-1644y-z+0=0.x+0.y+0.z+6302$</td><td>Br. xviii, 55.</td></tr>
<tr><td>5.</td><td>ka 6 ka 5 ka 2 ka 3</td><td>$\sqrt{6}+\sqrt{5}+\sqrt{2}+\sqrt{3}$</td><td>V. 37.</td></tr>
<tr><td>6.</td><td><table><tr><td>1</td><td>1</td><td>1</td><td>1</td></tr><tr><td>4</td><td>3</td><td>6</td><td>12</td></tr></table></td><td>$\frac{1}{4}+\frac{1}{3}+\frac{1}{6}+\frac{1}{12}$</td><td>S. 7.</td></tr>
<tr><td>7.</td><td><table><tr><td>1 1</td><td>1 1 1</td><td>1 1 1 1</td></tr><tr><td>1 2</td><td>1 2 3</td><td>1 2 3 5</td></tr></table></td><td>$1\times\frac{1}{2}+1\times\frac{1}{2}\times\frac{1}{3}+1\times\frac{1}{2}\times\frac{1}{3}\times\frac{1}{5}$</td><td></td></tr>
<tr><td>8.</td><td><table><tr><td>10</td><td>163</td><td>4</td><td>pha 163</td></tr><tr><td>1</td><td>60</td><td>1</td><td>150</td></tr></table></td><td>$10:\frac{163}{66}::4:\frac{163}{150}$</td><td>Bk. 27.</td></tr>
<tr><td>9.</td><td>1
1
1
$\dot{2}$
2
$\dot{9}$
1
$\dot{4}$
6
$\dot{1}0$

10.   40
1
1
3 +
1
4 +
1
5 +</td><td>9. $(1-\frac{1}{2})\ (1-\frac{2}{9})\ (1-\frac{1}{4})\ (1-\frac{6}{10})$
10. $40\ (1-\frac{1}{3})\ (1-\frac{1}{4})\ (1-\frac{1}{5})$</td><td>L. 53.
Bk. 25.</td></tr>
</table>

VI.

22. According to the Hindus the modern place-value system of arithmetical notation is of divine origin. This led the early orientalists to believe that, at any rate, the system had been in use in India from time immemorial; but an examination of the facts shows that the early notations in use were not place-value ones and that the modern place-value system was not introduced until comparatively modern times. The early systems employed may be conveniently termed (*a*) the Kharoshṭhī, (*b*) the Brāhmī, (*c*) Āryabhaṭa's alphabetic notation, (*d*) the word-symbol notation.

(*a*) The *Kharoshṭhī* script is written from right to left and was in use in the north-west of India and Central Asia at the beginning of the Christian era. The notation is shown in the accompanying table. It was, apparently, derived from the Aramaic system and has little direct connection with the other Indian notations. The smaller elements are written on the left.

(*b*) The *Brāhmī* notation is the most important of the old notations of India. It might appropriately be termed *the* Indian notation for it occurs in early inscriptions and was in fairly common use throughout India for many centuries, and even to the present day is occasionally used. The symbols employed varied somewhat in form according to time and place, but on the whole the consistency of form exhibited is remarkable. They are written from left to right with the smaller elements on the right. Several false theories as to the origin of these symbols have been published, some of which still continue to be recorded. The earliest orientalists gave them place-value, but this error soon disproved itself; it was then suggested that they were initial letters of numerical words; then it was propounded that the symbols were *aksharas* or syllables; then it was again claimed that the symbols were initial letters (this time *Kharoshṭhī*) of the corresponding numerals. These theories have been severally disproved.

The notation was possibly developed on different principles at different times. The first three symbols are natural and only differ from those of many other systems in consisting of horizontal instead of vertical strokes. No principle of formation of the symbols for "four" to "thirty" is now evident but possibly the "forty" was formed from the thirty by the addition of a stroke and the "sixty" and "seventy" and "eighty" and "ninety" appear to be connected in this way. The hundreds are (to a limited extent) evidently built upon such a plan, which, as BAYLEY pointed out, is the same as that employed in the Egyptian hieratic forms; but after the "three hundred" the Indian system forms the "four hundred" from the elements of "a hundred" and "four," and so on. The notation is exhibited in the table annexed.

PLACE VALUE NOTATIONS

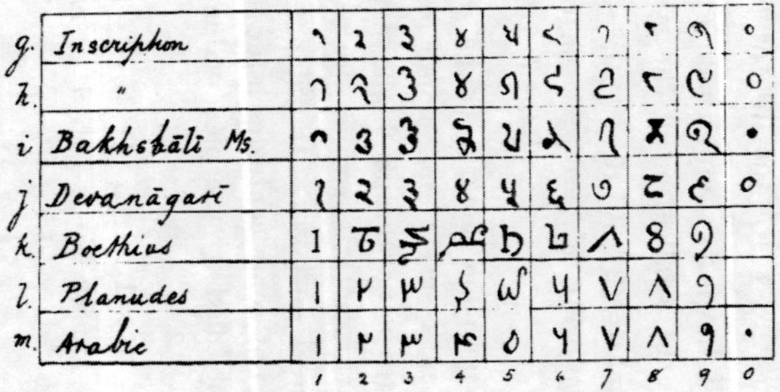

REFERENCES

a. After Bühler c. 100 A.D.

b. Inscriptions c. 100 – 500 A.D.

c. After Rapson c. 200 – 400 A.D

d. Bower MS. & Nepal MSS. c. 500 – 900 A.D.

e. After Bendall XIth century

f. Eastern Turkestan M.S. (Stein collection) J.R.A.S 1911, 455

g. Gwalior stone insc. ? IXth Cent. Arch. Surv. 1903-4 Pl LXXII

h. Surat grant of 1050 A.D. Ind. Ant. XII, 202. XI – 82.

NUMERICAL NOTATIONS

(c) Āryabhaṭa's alphabetic notation also had no place-value and differed from the Brāhmī notation in having the smaller elements on the left. It was, of course, written and read from left to right. It may be exhibited thus :

Letters ..	*k*	*kh*	*g*	*gh*	*ṅ*	*c*	*ch*	*j*	*jh*	*ñ*
Values ..	1	2	3	4	5	6	7	8	9	10.

Letters ..	*ṭ*	*ṭh*	*ḍ*	*ḍh*	*ṇ*	*t*	*th*	*d*	*dh*	*n*
Values ..	11	12	13	14	15	16	17	18	19	20.

Letters ..	*p*	*ph*	*b*	*bh*	*m*	*y*	*r*	*l*	*v*	*ś*	*sh*	*s*	*h*
Values ..	21	22	23	24	25	30	40	50	60	70	80	90	100.

The vowels indicate multiplication by powers of one hundred. The first vowel *a* may be considered as equivalent to 100^0, the second vowel $i=100^1$ and so on. The values of the vowels may therefore be shown thus :

Vowels ..	*a*	*i*	*u*	*ṛi*	*ḷi*	*e*	*ai*	*o*	*au*
Values ..	1	10^2	10^4	10^6	10^8	10^{10}	10^{12}	10^{14}	10^{16}

The following examples taken from Āryabhaṭa's *Gītikā* illustrate the application of the system :

khyughri$=(2+30).10^4+4.10^6=4320000$

cayagiyiṅuśulchḷi$=6+30+3.10^2+30.10^2+5.10^4+70.10^4$

$(50+7).10^8=57,753,336$

The notation could thus be used for expressing large numbers in a sort of mnemonic form. The table of sines referred to in paragraph 9 above was expressed by Aryabhaṭa in this notation which, by the way, he uses only for astronomical purposes. It did not come into ordinary use in India, but some centuries later it appears occasionally in a form modified by the place-value idea with the following values :

1	2	3	4	5	6	7	8	9	0
k	*kh*	*g*	*gh*	*ṅ*	*c*	*ch*	*j*	*jh*	*ñ*
ṭ	*ṭh*	*ḍ*	*ḍh*	*ṇ*	*t*	*th*	*d*	*dh*	*n*
p	*ph*	*b*	*bh*	*m*					
y	*r*	*l*	*v*	*ś*	*sh*	*s*	*h*	*l*	

Initial vowels are sometimes used as ciphers also. The earliest example of this modified system is of the twelfth century. Slight variations occur.

(*d*) *The word-symbol notation.*—A notation that became extraordinarily popular in India and is still in use was introduced about the ninth century, possibly from the East. In this notation any word that connotes the idea of a number may be used to denote that number : *e.g.* Two may be expressed by *nayana,* the eyes, or *karṇa,* the ears, etc. ; seven by *aśva,* the horses (of the sun) ; fifteen by *tithi,* the lunar days (of the half month) ; twenty by *nakha,* the nails (of the hands and feet) ; twenty-seven by *nakshatra* the lunar mansions ; thirty-two by *danta,* the teeth ; etc., etc.

(*e*) *The modern place-value notation.*—The orthodox view is that the modern place-value notation that is now universal was invented in India and until recently it was thought to have been in use in India at a very early date. Hindu tradition ascribes the invention to God ! According to Maçoudi a congress of sages, gathered together by order of king Brahma (who reigned 366 years), invented the nine figures ! An inscription of A.D. 595 is supposed to contain a genuine example of the system. According to M. NAU, the "Indian figures" were known in Syria in A.D. 662 ; but his authority makes such erroneous statements about "Indian" astronomy that we have no faith in what he says about other "Indian" matters. Certain other mediæval works refer to 'Indian numbers' and so on.

On the other hand it is held that there is no sound evidence of the employment in India of a place-value system earlier than about the ninth century. The suggestion of 'divine origin' indicates nothing but historical ignorance ; Maçoudi is obviously wildly erratic ; the inscription of A.D. 595 is not above suspicion* and the next inscription with an example of the place-value system is nearly three centuries

* The figures were obviously added at a later date.

later, while there are hundreds intervening with examples of the old non-place value system. The references in mediæval works to India do not necessarily indicate India proper but often simply refer to 'the East' and the use of the term with regard to numbers has been further confused by the misreading by Wœpcke and others of the Arabic term *hindasi* (geometrical, having to do with numeration, etc.) which has nothing to do with India. Again, it has been assumed that the use of the abacus "has been universal in India from time immemorial," but this assumption is not based upon fact, there being actually no evidence of its use in India until quite modern times. Further, there is evidence that indicates that the notation was introduced into India, as it was into Europe, from a right to left script.

23. In paragraph 7 above certain attempts at squaring the circle are briefly described and it has been pointed out (in § 10) that Āryabhaṭa gives an extremely accurate value of π. The topic is perhaps of sufficient interest to deserve some special mention. The Indian values given and used are not altogether consistent and the subject is wrapped in some mystery. Briefly put—the Indians record an extremely accurate value at a very early date but seldom or never actually use it. The following table roughly exhibits how the matter stands :—

Date Circa.	Authority.	Value of π
B.C. 1700	Ahmes the Egyptian	$(\frac{16}{9})^2 = 3{\cdot}160.$
,, 250	Archimedes	$< 3\frac{1}{7}$ and $> 3\frac{10}{71}$
,, ?	The Śulvasūtras	$(\frac{26}{15})^2 = 3{\cdot}004.$
	,,	$(\frac{9785}{5568})^2 = 3{\cdot}097.$
,, 230	Apollonius	$3\frac{17}{120} = 3{\cdot}14166.$
,, 120	Heron	$3 = 3{\cdot}$
	,,	$3\frac{1}{7} = 3{\cdot}14286.$

Date Circa.	Authority.	Value of π	
A.D. 150	Ptolemy	$3\frac{17}{120}$	$=3{\cdot}14166.$
,, 263	Liu Hiu	$3\frac{7}{50}$	$=3{\cdot}14.$
,, ?	Puliśa	$3\frac{177}{1250}$	$=3{\cdot}1416.$
,, 450	Tsu Ch'ung-chi	$3\frac{1}{7}$	$=3{\cdot}14286.$
,, 500	Aryabhaṭa	3	$=3{\cdot}$
	,,	$\frac{62832}{20000}$	$=3{\cdot}1416.$
	,,	$\frac{3393}{1080}$	$=3{\cdot}14166.$
,, 628	Brahmagupta	3	$=3{\cdot}$
	,,	$\sqrt{10}=3\frac{1}{7}$	$=3{\cdot}14286.$
	,,	$\sqrt{10}=\frac{721}{228}$	$=3{\cdot}16228.$
,, 800	M. ibn Musa	$\sqrt{10}$	$=?$
,, ?	Māhavīra	3	$=3{\cdot}$
	,,	$\sqrt{10}$	$=?$
,, 1020	Srīdhara	3	$=3{\cdot}$
	,,	$\sqrt{10}$	$=?$
	,,	$3\frac{1}{6}$	$=3{\cdot}1666.$
,, 1150	Bhāskara	3	$=3{\cdot}$
	,,	$3\frac{1}{7}$	$=3{\cdot}14286.$
	,,	$3\frac{17}{120}$	$=3{\cdot}14166.$
	,,	$3\frac{177}{1250}$	$=3{\cdot}1416.$
	Approximately correct value ..		$3{\cdot}14159.$

24. The mistakes made by the early orientalists have naturally misled the historians of mathematics, and the opinions of CHASLES, WŒPCKE, HANKEL and others founded upon such mistakes are now no longer authoritative. In spite, however, of the progress made in historical research there are still many errors current, of which, besides those already touched upon, the following may be cited as examples: (*a*) The proof by " casting out nines " is not of Indian origin and occurs in no Indian work before the 12th century; (*b*) The scheme of multiplication, of which the following is an Indian example of the 16th century, was known much earlier to the Arabs and there is no evidence that it is of Indian origin; (*c*) The *Regula duorum falsorum* occurs in no Indian work; (*d*) The Indians were not the first to give double solutions of quadratic equations; Bhaskāra was not the discoverer of the " principle of the differential calculus," etc., etc.

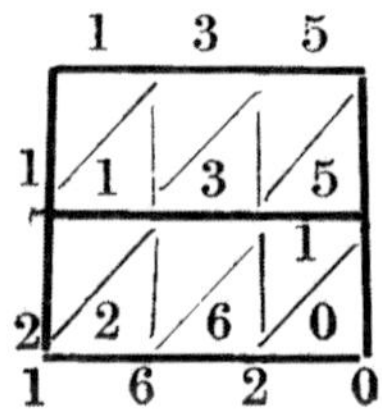

VII.

25. Of the personalities of the Indian mathematicians we know very little indeed but Alberuni has handed down Brahmagupta's opinion of Āryabhata and Puliśa* and his own opinions are worth repeating. We have also Bhāskara's inscription. The following notes contain, perhaps, all that is worth recording.

Alberuni writes (i,376) : 'Now it is evident that that which Brahmagupta relates on his own authority, and with which he himself agrees, is entirely unfounded ; but he is blind to this from sheer hatred of Āryabhaṭa, whom he abuses excessively. And in this respect Āryabhaṭa and Puliśa are the same to him. I take for witness the passage of Brahmagupta where he says that Āryabhaṭa has subtracted something from the cycles of *Caput Draconis* and of the *apsis* of the moon and thereby confused the computation of the eclipse. He is rude enough to compare Āryabhaṭa to a worm which, eating the wood, by chance describes certain characters in it, without intending to draw them. "He, however, who knows these things thoroughly stands opposite to Āryabhaṭa, Srīshena and Vishnuchandra like the lion against gazelles. They are not capable of letting him see their faces." In such offensive terms he attacks Āryabhaṭa and maltreats him.'

Again : '" Āryabhaṭa...... differs from the doctrine of the book *Smriti*, just mentioned, and he who differs from us is an opponent." On the other hand, Brahmagupta praises Puliśa for what he does, since he does not differ from the book *Smriti*.' Again, speaking of Varāhamihira, Srīshena,

* According to Alberuni Pulis'a was an Indian and Paulis'a a Greek.

Āryabhaṭa and Vishnuchandra, Brahmagupta says: 'If a man declares these things illusory he stands outside the generally acknowledged dogma, and that is not allowed.'

Of Varāhamihira, Alberuni writes: 'In former times, the Hindus used to acknowledge that the progress of science due to the Greeks is much more important than that which is due to themselves. But from this passage of Varāhamihira alone (*see* paragraph 2 above) you see what a self-lauding man he is, whilst he gives himself airs as doing justice to others;' but in another place (ii, 110) Alberuni says: 'On the whole his foot stands firmly on the basis of truth and he clearly speaks out the truth.....Would to God all distinguished men followed his example.'

Of Brahmagupta, Alberuni writes (ii, 110): 'But look, for instance, at Brahmagupta, who is certainly the most distinguished of their astronomers....he shirks the truth and lends his support to imposture......under the compulsion of some mental derangement, like a man whom death is about to rob of his consciousness.....If Brahmagupta....is one of those of whom God says, "They have denied our signs, although their hearts knew them clearly, from wickedness and haughtiness," we shall not argue with him, but only whisper into his ear—"If people must under circumstances give up opposing the religious codes (as seems to be your case), why then do you order people to be pious if you forget to be so yourself".....I, for my part, am inclined to the belief that that which made Brahmagupta speak the above mentioned words (which involve a sin against conscience) was something of a calamitous fate, like that of Socrates, which had befallen him, notwithstanding the abundance of his knowledge and the sharpness of his intellect, and notwithstanding his extreme youth at the time. For he wrote the *Brahmasiddhānta* when he was only thirty years of age. If this indeed is his excuse we accept it and drop the matter.'

An inscription found in a ruined temple at Pāṭṇā, a deserted village of Khandesh in the Bombay Presidency, refers to Bhāskara in the following terms : 'Triumphant is the illustrious Bhāskarāchārya whose feet are revered by the wise, eminently learned....who laid down the law in metrics, was deeply versed in the Vaiseshika system,....was in poetics a poet, like unto the three-eyed in the three branches, the multifarious arithmetic and the rest....Bhāskara, the learned, endowed with good fame and religious merit, the root of the creeper—true knowledge of the Veda, an omniscient seat of learning ; whose feet were revered by crowds of poets, etc.

The inscription goes on to tell us of Bhāskara's grandson 'Changadeva, chief astrologer of King Simghana, who, to spread the doctrines promulgated by the illustrious Bhāskarāchārya, founds a college, that in his college the *Siddhānta-śiromaṇi* and other works composed by Bhāskara, as well as other works by members of his family, shall be necessarily expounded.'

Bhāskara's most popular work is entitled the *Līlāvatī* which means 'charming.' He uses the phrase "Dear intelligent *Līlāvatī*," etc., and thus have arisen certain legends as to a daughter he is supposed to be addressing. The legends have no historical basis.

Bhāskara at the end of his *Vīja gaṇita* refers to the treatises on algebra of Brahmagupta, Śrīdhara and Padmanābha as "too diffusive" and states that he has compressed the substance of them in "a well reasoned compendium, for the gratification of learners."

VIII.

26. Chinese Mathematics.—There appears to be abundant evidence of an intimate connection between Indian and Chinese mathematics. A number of Indian embassies to China and Chinese visits to India are recorded in the fourth and succeeding centuries. The records of these visits are not generally found in Indian works and our knowledge of them in most cases comes from Chinese authorities, and there is no record in Indian works that would lead us to suppose that the Hindus were in any way indebted to China for mathematical knowledge. But, as pointed out before, this silence on the part of the Hindus is characteristic, and must on no account be taken as an indication of lack of influence. We have now before us a fairly complete account of Chinese mathematics* which appears to prove a very close connection between the two countries. This connection is briefly illustrated in the following notes.

The earliest Chinese work that deals with mathematical questions is said to be of the 12th century B.C. and it records an acquaintance with the Pythagorean theorem. Perhaps the most celebrated Chinese mathematical work is the *Chin-chang Suan-shū* or "Arithmetic in Nine Sections" which was composed at least as early as the second century B.C. while Chang T'sang's commentary on it is known to have been written in A.D. 263. The "Nine Sections" is far more complete than any Indian work prior to Brahmagupta (A.D. 628) and in some respects is in advance of that writer. It treats of fractions, percentage, partnership, extraction of square and cube-roots, mensuration of plane figures and solids, problems involving equations of the first and second

* By Yoshio Mikami.

degree. Of particular interest to us are the following : The area of a segment of a circle$=\frac{1}{2}(c+a)a$, where 'c' is the chord and 'a' the perpendicular, which actually occurs in Mahāvīra's work ; in the problems dealing with the evaluation of roots, partial fractions with unit numerators are used (*cf.* paragraphs 5 and 7 above) ; the diameter of a sphere$=\sqrt[3]{\frac{16}{9}\times\text{volume}}$, which possibly accounts for Āryabhata's strange rule ; the volume of the cone$=\left(\frac{\text{circumference}}{6}\right)^2$ which is given by all the Indians ; and the correct volume for a truncated pyramid which is reproduced by Brahmagupta and Srīdhara. One section deals with right angled triangles and gives a number of problems like the following :

"There is a bamboo 10 feet high, the upper end of which being broken reaches to the ground 3 feet from the stem. What is the height of the break ?" This occurs in every Indian work after the 6th century. The problem about two travellers meeting on the hypotenuse of a right-angled triangle occurs some ten centuries later in exactly the same form in Mahāvīra's work.

The *Sun-Tsū Suan-ching* is an arithmetical treatise of about the first century. It indulges in big numbers and elaborate tables like those contained in Mahāvīra's work ; it gives a clear explanation of square-root and it contains examples of indeterminate equations of the first degree. The example : "There are certain things whose number is unknown. Repeatedly divided by 3 the remainder is 2 ; by 5 the remainder is 3, and by 7 the remainder is 2. What will be the number ?" re-appears in Indian works of the 7th and 9th centuries. The earliest Indian example is given by Brahmagupta and is : "What number divided by 6 has a remainder 5, and divided by 5 has a remainder of 4 and by 4 a remainder of 3, and by 3 a remainder of 2 ?" Mahāvīra has similar examples.

In the 3rd century the *Sea Island Arithmetical Classic* was written. Its distinctive problems concern the measurement of the distance of an island from the shore, and the solution given occurs in Āryabhaṭa's *Gaṇita* some two centuries later. The *Wu-t'sao* written before the 6th century appears to indicate some deterioration. It contains the erroneous rule for areas given by Brahmagupta and Mahāvīra. The arithmetic of *Chang-Ch'iu-chien* written in the 6th century contains a great deal of matter that may have been the basis of the later Indian works. Indeed the later Indian works seem to bear a much closer resemblance to Chang's arithmetic than they do to any earlier Indian work.

The problem of " the hundred hens " is of considerable interest. Chang gives the following example : " A cock costs 5 pieces of money, a hen 3 pieces and 3 chickens 1 piece. If then we buy with 100 pieces 100 of them what will be their respective numbers ?"

No mention of this problem is made by Brahmagupta, but it occurs in Mahāvīra and Bhāskara in the following form : " Five doves are to be had for 3 *drammas*, " 7 cranes for 5, 9 geese for 7 and 3 peacocks for 9. Bring 100 of these birds for 100 *drammas* for the prince's gratification." It is noteworthy that this problem was also very fully treated by Abū Kāmil (Shogâ) in the 9th century, and in Europe in the middle ages it acquired considerable celebrity.

Enough has been said to show that there existed a very considerable intimacy between the mathematics of the Indians and Chinese ; and assuming that the chronology is roughly correct, the distinct priority of the Chinese mathematics is established. On the other hand Brahmagupta gives more advanced developments of indeterminate equations than occurs in the Chinese works of his period, and it is not until after Bhāskara that Ch'in Chu-sheo recorded (in A.D. 1247) the celebrated *t'ai-yen ch'in-yi-shu* or process of indeterminate analysis, which is, however, attributed to I'-hsing nearly six centuries earlier. The Chinese had maintained intellectual intercourse with India since the first century and had translated many Indian (Buddhistic) works. They (unlike their Indian friends) generally give the source of their information and acknowledge their indebtedness with becoming courtesy. From the 7th century Indian scholars were occasionally employed on the Chinese Astronomical Board. Mr. Yoshio Mikami states that there is no evidence of Indian influence on Chinese mathematics. On the other hand he says " the discoveries made in China may have touched the eyes of Hindoo scholars."

27. Arabic Mathematics.—It has often been assumed, with very little justification, that the Arabs owed their knowledge of mathematics to the Hindus.

Muhammad b. Mūsā el-Chowārezmi (A.D. 782) is the earliest Arabic writer on mathematics of note and his best known work is the *Algebra.* The early orientalists appear to have been somewhat prejudiced against Arabic scholarship for, apparently without examination, they ascribed an Indian origin to M. b. Mūsā's work. The argument used was as follows : " There is nothing in history," wrote Cossali, and Colebrooke repeated it, ' respecting Muhammad ben Mūsā individually, which favours the opinion that he took from the Greeks, the algebra which he taught to the Muhammadans. History presents him in no other light than a mathematician of a country most distant from Greece and contiguous to India.....Not having taken algebra from the Greeks, he must either have invented it himself or taken it from the Indians.' As a matter of fact his algebra shows, as pointed out by Rodet, no sign of Indian influence and is practically wholly based upon Greek knowledge ; and it is now well known that the development of mathematics among the Arabs was largely, if not wholly, independent of Indian influence and that, on the other hand, Indian writers on mathematics later than Brahmagupta were possibly influenced considerably by the Arabs. Alberuni early in the 11th century wrote : ' You mostly find that even the so-called scientific theorems of the Hindus are in a state of utter confusion, devoid of any logical order....since they cannot raise themselves to the methods of strictly scientific deduction....I began to show them the elements on which this science rests, to point out to them some rules of logical deduction and the scientific method of all mathematics, etc.'

The fact is that in the time of el-Māmūn (A.D. 772) a certain Indian astronomical work (or certain works) was translated into Arabic. On this basis it was *assumed* that the Arabic astronomy and mathematics was wholly of Indian origin, while the fact that Indian works were translated is really only evidence of the intellectual spirit then prevailing in Baghdad. No one can deny that Āryabhaṭa and Brahmagupta preceded M. b. Mūsā* but the fact remains that there is not the slightest resemblance between the previous Indian works and those of M. b. Mūsā. The point was somewhat obscured by the publication in Europe of an arithmetical treatise by M. b. Mūsā under the title *Algoritmi de Numero Indorum*. As is well known the term India did not in mediæval times necessarily denote the India of to-day and despite the title there is nothing really Indian in the work. Indeed its contents prove conclusively that it is not of Indian origin. The same remarks apply to several other mediæval works.

28. From the time of M. b. Mūsā onwards the Muhammadan mathematicians made remarkable progress. To illustrate this fact we need only mention a few of their distinguished writers and their works on mathematics. 'Tābit b. Qorra b. Merwān (826-901) wrote on Euclid, the Almagest, the arithmetic of Nicomachus, the right-angle triangle the parabola, magic squares, amicable numbers, etc. Qosṭā b. Lūka el-Ba'albekī (died c. A.D. 912) translated Diophantus

---

* It should not be forgotten, however, that Nicomachus (A.D. 100) was an Arabian, while Jamblichus, Damascius, and Eutocius were natives of Syria.

and wrote on the sphere and cylinder, the rule of two errors, etc. El-Battānī (M. b. Gābir b. Sinān, A.D. 877-919) wrote a commentary on Ptolemy and made notable advances in trigonometry. Abū Kāmil Shogā b. Aslam (c. 850-930) wrote on algebra and geometry, the pentagon and decagon, the rule of two errors, etc. Abū 'l-Wefā el-Būzgāni, born in A.D. 940, wrote commentaries on Euclid, Diophantus, Hipparchus, and M. b. Mūsā, works on arithmetic, on the circle and sphere, etc., etc. Abū Sa 'īd, el-sigzī (Ahmed b. M. b. Abdelgalil, A.D., 951-1024) wrote on the trisection of an angle, the sphere, the intersection of the parabola and hyperbola, the Lemmata of Archimedes, conic sections, the hyperbola and its asymptotes, etc., etc. Abū Bekr, el-Karchi (M. b. el-Hasan, 1016 A.D.) wrote on arithmetic and indeterminate equations after Diophantus. Alberuni (M. b. Aḥmed, Abū'l-Riḥān el-Bīrūnī) was born in A.D. 973 and besides works on history, geography, chronology and astronomy wrote on mathematics generally, and in particular on tangents, the chords of the circle, etc. Omār b. Ibrāhīm el-Chaijāmi, the celebrated poet, was born about A.D. 1046 and died in A.D. 1123 a few years after Bhāskara was born. He wrote an algebra in which he deals with cubic equations, a commentary on the difficulties in the postulates of Euclid ; on mixtures of metals ; and on arithmetical difficulties.

This very brief and incomplete resumé of Arabic mathematical works written during the period intervening between the time of Brahmagupta and Bhāskara indicates at least considerable intellectual activity and a great advance on the Indian works of the period in all branches of mathematics except, perhaps, indeterminate equations.

IX.

29. That the most important parts of the works of the Indian mathematicians from Āryabhaṭa to Bhāskara are essentially based upon western knowledge is now established. A somewhat intimate connection between early Chinese and Indian mathematics is also established—but the connection in this direction is not very intimate with respect to those sections that may be termed Greek, *e.g.*, quadratic indeterminates, cyclic quadrilaterals, etc. That the Arabic development of mathematics was practically independent of Indian influence is also proved.

The Arab mathematicians based their work almost wholly upon Greek knowledge; but the earliest of them known to us, M. b. Mūsā, flourished after Brahmagupta so that the Arabs could not have been the intermediaries between the Greeks and Indians. Indeed their chronological position has misled certain writers to the erroneous conclusion that they obtained their elements of mathematics from the Indians.

Other possible paths of communication between the Indians and Greeks are by way of China and by way of Persia. The former is not so improbable as it at first seems. Further information about the early silk trade with China might possibly throw light on the subject. The intellectual communication between India and China at the critical period is well known—there being numerous references to such communication in Chinese literature. If sound translations of the early Chinese mathematical works were available we might be able to draw more definite conclusions, but as the evidence now stands there is nothing that would warrant more than the bare suggestion of a Chinese source.

We have already mentioned the visit of certain Greek mathematicians to the Court of Chosroes I, and there are certain other facts which at least justify the consideration of the Persian route. The Sássánid period, A.D. 229-652, shows a somewhat remarkable parallelism with the age of enlightenment in India that roughly corresponds with the Gupta period. "The real missionaries of culture in the Persian empire at this time were the Syrians, who were connected with the west by their religion and who, in their translations, diffused Greek literature throughout the orient." Mr. Vincent Smith discusses the probability of Sássánian influence on India but states that there is no direct evidence.

Although it may be possible to offer only conjectures as to the actual route by which any particular class of Greek knowledge reached India, the fact remains that during the period under consideration the intellectual influence of Greek on India was considerable. It is evident not only in the mathematical work of the Indians but also in sculpture, architecture, coinage, astronomy, astrology, &c. Mr. Vincent Smith refers "to the cumulative proof that the remarkable intellectual and artistic output of the Gupta period was produced in large measure by reason of the contact between the civilization of India and that of the Roman Empire;" and research is almost daily adding to such proof.

The flourishing state of the Gupta empire, the greatest in India since the days of Asoka, and the wise influence of its principal rulers gave a great impetus to scholarship of all kinds. The numerous embassies to and from foreign countries—which were means of intellectual as well as political communication—no doubt contributed to the same end; and the knowledge of Greek works displayed by Āryabhaṭa, Varāha Mihira, and Brahmagupta was one of the natural results of this renaissance of learning.

APPENDIX I.

Extracts from Texts.

The Śulvasūtras.

*1. In the following we shall treat of the different manners of building the *agni*. 2. We shall explain how to measure out the circuit of the area required for them.

* * * * *

45. The cord stretched across a square produces an area of twice the size.

46. Take the measure for the breadth, the diagonal of its square for the length : the diagonal of that oblong is the side of a square the area of which is three times the area of the square.

* * * * *

48. The diagonal of an oblong produces by itself both the areas which the two sides of the oblong produce separately.

49. This is seen in those oblongs whose sides are three and four, twelve and five, fifteen and eight, seven and twenty-four, twelve and thirty-five, fifteen and thirty-six.

* * * * *

51. If you wish to deduct one square from another square cut off a piece from the larger square by making a mark on the ground with the side of the smaller square which you wish to deduct. Draw one of the sides across the oblong so that it touches the other side. Where it touches there cut off. By this line which has been cut off the small square is deducted from the large one.

* * * * *

---

* These numbers refer to Baudhāyana's edition as translated by Dr. Thibaut.

58. If you wish to turn a square into a circle draw half of the cord stretched in the diagonal from the centre towards the *prāchī* line. Describe the circle together with the third part of that piece of the cord which stands over.

Āryabhaṭa's gaṇita—(Circa. A.D. 520).

6. The area of a triangle is the product of the perpendicular common to the two halves and half the base. Half the product of this and the height is the solid with six edges.

* * * * *

10. Add four to one hundred, multiply by eight and add again sixty-two thousand. The result is the approximate value of the circumference when the diameter is twenty thousand.

* * * * *

13. The circle is produced by a rotation ; the triangle and the quadrilateral are determined by their hypotenuses ; the horizontal by water and the vertical by the plumb line.

* * * * *

29. The sum of a certain number of terms diminished by each term in succession added to the whole and divided by the number of terms less one gives the value of the whole.

Brahmagupta—(Born A.D. 598).

1. He who distinctly knows addition and the rest of the twenty operations and the eight processes including measurement by shadows is a mathematician.

* * * * *

14. The principal multiplied by its time and divided by the interest, and the quotient being multiplied by the factor less one is the time. The sum of principal and interest divided by unity added to the interest on unity is the principal.

* * * * *

17. The number of terms less one multiplied by the common difference and added to the first term is the amount of the last. Half the sum of the last and first terms is the mean amount, and this multiplied by the number of terms is the sum of the whole.

* * * * *

21. The product of half the sides and opposite sides is the rough area of a triangle or quadrilateral. Half the sum of the sides set down four times and each lessened by the sides being multiplied together—the square-root of the product is the exact area.

* * * * *

40. The diameter and the square of the radius respectively multiplied by three are the practical circumference and area. The square-roots extracted from ten times the square of the same are the exact values.

* * * * *

62. The integer multiplied by the sexagesimal parts of its fraction and divided by thirty is the square of the minutes and is to be added to the square of the whole degrees.

* * * * *

101. These questions are stated merely for gratification. The proficient may devise a thousand others or may solve by the rules taught problems set by others.

102. As the sun obscures the stars so does the expert eclipse the glory of other astronomers in an assembly of people by reciting algebraic problems, and still more by their solution.

Mahāvīra's Gaṇita-Sāra-Sangraha—(Circa. A.D. 850).

i. 13–14. The number, the diameter and the circumference of islands, oceans and mountains ; the extensive dimensions of the rows of habitations and halls belonging to the inhabitants of the world, of the interspace, of the world of light, of the world of the gods and to the dwellers in hell, and miscellaneous measurements of all sorts—all these are made out by means of computation.

* * * * *

vi. 147. Divide by their rate prices. Diminish by the least among them and then multiply by the least the mixed price of all the things and subtract from the given number of things. Now split up (this) into as many (as there are left) and then divide. These separated from the total price give the price of the dearest article of purchase. [This is a solution of example 36 below.]

* * * * *

vi. 169. It has to be known that the products of gold as multiplied by their colours when divided by the mixed gold gives rise to the resulting colour (*varṇa*). [See examples 24 and 25 below.]

* * * * *

vii. 2. Area has been taken to be of two kinds by Jina in accordance with the result—namely, that which is for practical purposes and that which is minutely accurate.

* * * * *

vii. 233. Thus ends the section of devilishly difficult problems.

Srīdhara's Triśatikā—(Circa A.D. 1030).

1. Of a series of numbers beginning with unity and increasing by one, the sum is equal to half the product of the number of terms and the number of terms together with unity.

* * * * *

32. In exchange of commodities the prices being transposed apply the previous rule (of three). With reference to the sale of living beings the price is inversely proportional to their age.

* * * * *

65. If the gnomon be divided by twice the sum of the gnomon and the shadow the fraction of the day elapsed or which remains will be obtained.

Bhāskara—(Born A.D. 1114).

L. 1. I propound this easy process of calculation, delightful by its elegance, perspicuous with concise terms, soft and correct and pleasing to the learned.

* * * * *

L. 139. A side is put. From that multiplied by twice some assumed number and divided by one less than the square of the assumed number a perpendicular is obtained. This being set aside is multiplied by the arbitrary number and the side as put is subtracted—the remainder will be the hypotenuse. Such a triangle is termed 'genuine.'

* * * * *

L. 189. Thus, with the same sides, may be many diagonals in the quadrilateral. Yet, though indeterminate, diagonals have been sought as determinate by Brahmagupta and others.

* * * * *

L. 213. The circumference less the arc being multiplied by the arc the product is termed 'first.' From the quarter of the square of the circumference multiplied by five subtract that first product. By the remainder divide the first product multiplied by four times the diameter. The quotient will be the chord.

* * * * *

V. 170. In the like suppositions, when the operation, owing to restriction, disappoints the answer must by the intelligent be discovered by the exercise of ingenuity. Accordingly it is said The conditions—a clear intellect, assumption of unknown quantities, equation, and the rule of three—are means of operation in analysis.'

* * * * *

V. 224. The rule of three terms is arithmetic; spotless understanding is algebra. What is there unknown to the intelligent? Therefore for the dull alone it is set forth.

V. 225. To augment wisdom and strengthen confidence, read, read, mathematician, this abridgement elegant in style, easily understood by youth, comprising the whole essence of calculation and containing the demonstration of its principles—full of excellence and free from defect.

APPENDIX II.

EXAMPLES.

1. One-half, one-sixth, and one-twelfth parts of a pole are immersed respectively under water, clay, and sand. Two *hastas* are visible. Find the height of the pole ?

Answer—8 hastas. *S.* 23.

2. The quarter of a sixteenth of the fifth of three-quarters of two-thirds of half a *dramma* was given to a beggar by a person from whom he asked alms. Tell how many cowries the miser gave if thou be conversant in arithmetic with the reduction termed sub-division of fractions ?

Answer—1 cowrie. *L.* 32.
(1,280 cowries=1 dramma).

3. Out of a swarm of bees one-fifth settled on a blossom of kadamba, one-third on a flower of silīndhrī, three times the difference of those numbers flew to a bloom of kutaja. One bee, which remained, hovered and flew about in the air, allured at the same moment by the pleasing fragrance of a jasmine and pandanus. Tell me, charming woman, the number of bees ?

Answer—15. *L.* 54, *V.* 108.

4. The third part of a necklace of pearls broken in an amorous struggle fell on the ground. Its fifth part was seen resting on the couch, the sixth part was saved by the lady and the tenth part was taken up by her lover. Six pearls

---

* *L*=the Līlāvatī, *V*=Vīja Gaṇita, both by Bhāskara, *M*=Mahāvira, *S*=Srīdhara, *C*=Chaturveda.

remained on the string. Say, of how many pearls the necklace was composed ?

Answer—30. *S.* 26.

5. A powerful, unvanquished, excellent black snake which is 32 *hastas* in length enters into a hole $7\frac{1}{2}$ *añgulas* in $\frac{5}{14}$ of a day, and in the course of a quarter of a day its tail grows by $2\frac{3}{4}$ *añgulas*. O ornament of arithmeticians, tell me by what time this same enters fully into the hole ?

Answer—$76\frac{4}{5}$ days. *M. v*, 31.

(24 añgulas=1 hasta.)

6. A certain person travels at the rate of 9 *yojanas* a day and 100 *yojanas* have already been traversed. Now a messenger sent after goes at the rate of 13 *yojanas* a day. In how many days will he overtake the first person ?

Answer—25. *M. vi*, 327.

7. A white-ant advances 8 *yavas* less one-fifth in a day and returns the twentieth part of an *añgula* in 3 days. In what space of time will one, whose progress is governed by these rates of advancing and receding proceed 100 *yojanas* ?

Answer—98042553 days. *C.* 283.

(8 yavas=1 añgula, 768000 añgulas=1 yojana).

8. Twenty men have to carry a palanquin two *yojanas* and 720 *dīnāras* for their wages. Two men stop after going two *krośas*, after two more *krośas* three others give up, and after going half the remaining distance five men leave. What wages do they earn ?

Answers—18, 57, 155, 490. *M. vi*, 231.

(4 krośas=1 yojana).

9. It is well known that the horses belonging to the sun's chariot are seven. Four horses drag it along being

---

* *L*=the Līlāvatī, *V*=Vīja Gaṇita, both by Bhāskara, *M*=Mahāvira, *S*=Srīdhara, *C*=Chaturveda

harnessed to the yoke. They have to do a journey of 70 *yojanas*. How many times are they unyoked and how many times yoked in four ?

Answer—Every 10 yojanas, and each horse travels 40 yojanas. *M. VI*, 158.

10. If a female slave sixteen years of age brings thirty-two, what will one twenty cost ?

Answer—$25\frac{3}{5}$. *L.* 76.

11. Three hundred gold coins form the price of 9 damsels of 10 years. What is the price of 36 damsels of 16 years ?

Answer—750. *M. V*, 40.

12. The price of a hundred bricks, of which the length, thickness and breadth respectively are 16, 8 and 10, is settled at six *dīnāras*, we have received 100,000 of other bricks a quarter less in every dimension. Say, what we ought to pay ?

Answer—$2531\frac{1}{4}$. *C.* 285.

13. Two elephants, which are ten in length, nine in breadth, thirty-six in girth and seven in height, consume one *droṇa* of grain. How much will be the rations of ten other elephants which are a quarter more in height and other dimensions ?

Answer—12 droṇas, 3 prasthas, $1\frac{1}{2}$ kuḍavas. *C.* 285.
(64 kuḍavas=16 prasthas=1 droṇa).

14. One bestows alms on holy men in the third part of a day, another gives the same in half a day and a third distributes three in five days. In what time, keeping to these rates, will they have given a hundred ?

Answer—$17\frac{6}{7}$. *C.* 282.

---

* *L*=the Lilāvati, *V*=Vija Gaṇita, both by Bhāskara, *M*=Mahāvira *S*=Sridhara, *C*=Chaturveda.

15. Say, mathematician, what are the apportioned shares of three traders whose original capitals were respectively 51, 68 and 85, which have been raised by commerce conducted by them in joint stock to the aggregate amount of 300 ?

Answer—75, 100, 125. *L.* 93.

16. One purchases seven for two and sells six for three. Eighteen is the profit. What is the capital ?

Answer—$18 \div (\frac{3/6}{2/7} - 1) = 24$. *Bakhshāli Ms.* 54.

17. If a *pala* of best camphor may be had for two *nishkas*, and a *pala* of sandal wood for the eighth part of a *dramma* and half a *pala* of aloe wood also for the eighth of a *dramma*, good merchant, give me the value of one *nishka* in the proportions of 1, 16 and 8 ; for I wish to prepare a perfume ?

Answer—Prices—Drammas $14\frac{2}{9}$, $\frac{8}{9}$, $\frac{8}{9}+\frac{4}{9}$, $\frac{64}{9}$, $\frac{32}{9}$ *L.* 98.
(16 drammas=1 nishka).

18. If three and a half *mānas* of rice may be had for one *dramma* and eight of beans for the same price, take these thirteen *kākinīs*, merchant, and give me quickly two parts of rice and one of beans : for we must make a hasty meal and depart, since my companion will proceed onwards ?

Answer— $\frac{7}{12}$ and $\frac{7}{24}$. *L.* 97, *V.* 115.
(64 kākinīs=1 dramma).

19. If the interest on 200 for a month be 6 *drammas*, in what time will the same sum lent be tripled ?

Answer—$66\frac{2}{3}$ months. *C.* 287.

20. If the principal sum with interest at the rate of five on the hundred by the month amount in a year to one thousand, tell the principal and interest respectively ?

Answer—625, 375. *L.* 89.

---

* *L*=the Līlāvatī, *V*=Vīja Gaṇita, both by Bhāskara, *M*=Mahāvira, *S*=Srīdhara, *C*=Chaturveda.

21. In accordance with the rate of five per cent. (per mensem) two months is the time for each instalment ; and paying the instalments of 8 (on each occasion) a man became free in 60 months. What is the capital ?

Answer—60. *M. vi*, 64.

22. Five hundred *drammas* were a loan at a rate of interest not known. The interest of that money for four months was lent to another person at the same rate and it accumulated in ten months to 78. Tell the rate of interest on the principal ?

Answer—60. *C.* 288.

23. Subtracting from a sum lent at five in the hundred the square of the interest, the remainder was lent at ten in the hundred. The time of both loans was alike and the amount of interest equal ?

Answer—Principal 8. *V.* 109.

24. There is 1 part of 1 *varṇa*, 1 part of 2 *varṇas*, 1 part of 3 *varṇas*, 2 parts of 4 *varṇas*, 4 parts of 5 *varṇas*, 7 parts of 14 *varṇas*, and 8 parts of 15 *varṇas*. Throwing these into the fire make them all into one and then what is the *varṇa* of the mixed gold ?

Answer—$10\frac{1}{2}$. *M. vi*, 170.

[The term *varṇa* corresponds to 'carat' or measure of 'purity of gold.']

25. Gold 1, 2, 3, 4 *suvarnas*, and losses 1, 2, 3, 4 *māshakas*.

The average loss is $\frac{1.1+2.2+3.3+4.4}{1+2+3+4}=3$.

Bakhshāli Ms. 27.

* *L*=the Līlāvatī, *V*=Vīja Gaṇita, both by Bhāskara, *M*=Mahāvira, *S*=Srīdhara, *C*=Chaturveda.

26. Of two arithmetical progressions with equal sums and the same number of terms the first terms are 2 and 3, the increments 3 and 2 respectively and the sum 15. Find the number of terms ?

Answer—3. *Bakhshāli Ms.* 18.

27. A merchant pays octroi on certain goods at three different places. At the first he gives $\frac{1}{3}$ of the goods, at the second $\frac{1}{4}$, and at the third $\frac{1}{5}$. The total duty paid is 24. What was the original amount of the goods ?

Answer—40. *Bakhshāli Ms.* 25.

28. One says : "Give me a hundred and I shall be twice as rich as you, friend !" The other replies : "If you deliver ten to me I shall be six times as rich as you. Tell me what was the amount of their respective capitals ?

Answer—40 and 170. *V.* 106, 156.

29. *A* gives a certain amount, *B* gives twice as much as *A*, *C* gives 3 times as much as *B*, *D* gives 4 times as much as *C* and the total is 132.

Answer—*A* gives 4, etc. *Bakhshāli Ms.* 54.

30. Four jewellers possessing respectively 8 rubies, 10 sapphires, 100 pearls and 5 diamonds, presented each from his own stock one apiece to the rest in token of regard and gratification at meeting ; and thus they became owners of stock of precisely the same value. Tell me, friend, what were the prices of their gems respectively ?

Answer—24, 16, 1, 96 [These are relative values only]. *L.* 100.

31. The quantity of rubies without flaw, sapphires, and pearls belonging to one person, is five, eight and seven

* *L*=the Lilāvatī, *V*=Vīja Gaṇita, both by Bhāskara, *M*=Mahāvira, *S*=Srīdhara, *C*=Chaturveda.

respectively. The number of like gems belonging to another is seven, nine and six. One has ninety, the other sixty-two rupees. They are equally rich. Tell me quickly, then, intelligent friend, who art conversant with algebra, the prices of each sort of gem ?

Answer—14, 1, 1, etc. *V*. 105 & 156.

[Bhāskara ' assumes ' relative values.]

32. The horses belonging to these four persons respectively are five, three, six and eight ; the camels belonging to them are two, seven, four and one ; their mules are eight, two, one and three ; and the oxen owned by them are seven, one, two and one. All are equally rich. Tell me severally, friend, the rates of the prices of horses and the rest ?

Answer—85, 76, 31, 4, etc. *V*. 157.

33. Say quickly, friend, in what portion of a day will four fountains, being let loose together fill a cistern, which, if opened one by one, would fill it in one day, half a day, the third and the sixth parts respectively ?

Answer—$\frac{1}{12}$. *L*. 95.

34. The son of Pṛithā, angered in combat, shot a quiver of arrows to slay Karṇa. With half his arrows he parried those of his antagonist ; with four times the square-root of the quiverful he killed his horse ; with six arrows he slew Salya ; with three he demolished the umbrella, standard and bow ; and with one he cut off the head of the foe. How many were the arrows which Arjuna let fly ?

Answer—100. *L*. 67, *V*. 133.

35. For 3 *paṇas* 5 *palas* of ginger are obtained, for 4 *paṇas* 11 *palas* of long pepper and for 8 *paṇas* 1 *pala* of

---

* *L*=the Līlāvatī, *V*=Vīja Gaṇita, both by Bhāskara, *M*=Mahāvira, *S*=Srīdhara, *C*=Chaturveda.

pepper. By means of the purchase money of 60 *paṇas* quickly obtain 68 *palas*? *M. vi*, 150.

Answer—Ginger 20, long pepper 44, pepper 4.

36. Five doves are to be had for three *drammas*; seven cranes for five; nine geese for seven and three peacocks for nine. Bring a hundred of these birds for a hundred *drammas* for the prince's gratification?

V. 158-9; *M. vi*, 152.

Answer—Prices 3, 40, 21, 36.

Birds 5, 56, 27, 12.

(This class of problem was treated fully by Abú Kāmil-el-Misri (*c.* 900 *A.D.*). See H. Suter: *Das Buch der Seltenheit, etc.* Bibliotheca Mathematica 11 (1910-11), pp.100-120.

*   *   *   *   *

*   *   *   *   *

*   *   *   *   *

37. In a certain lake swarming with red geese the tip of a bud of a lotus was seen half a *hasta* above the surface of the water. Forced by the wind it gradually advanced and was submerged at a distance of two *hastas*. Calculate quickly, O mathematician, the depth of the water?

Answer—$\frac{15}{4}$. *L.* 153; *V.* 125.

38. If a bamboo measuring thirty-two *hastas* and standing upon level ground be broken in one place by the force of the wind and the tip of it meet the ground at sixteen *hastas*, say, mathematician, at how many *hastas* from the root it is broken?

Answer—12. *L.* 148.

---

* *L*=the Līlāvatī, *V*=Vīja Gaṇita, both by Bhāskara, *M*=Mahāvīra, *S*=Srīdhara, *C*=Chaturveda.

39. A snake's hole is at the foot of a pillar 9 *hastas* high and a peacock is perched on the summit. Seeing a snake, at a distance of thrice the pillar, gliding towards his hole, he pounces obliquely on him. Say quickly at how many *hastas* from the snake's hole do they meet, both proceeding an equal distance ?

Answer—12. *L.* 150.

40. From a tree a hundred *hastas* high, a monkey descended and went to a pond two hundred *hastas* distant, while another monkey, jumping a certain height off the tree, proceeded quickly diagonally to the same spot. If the space travelled by them be equal, tell me quickly, learned man, the height of the leap, if thou hast diligently studied calculation ?

Answer—50. *L.* 155 ; *V.* 126.

41. The man who travels to the east moves at the rate of 2 *yojanas,* and the other man who travels northward moves at the rate of 3 *yojanas.* The latter having journeyed for 5 days turns to move along the hypotenuse. In how many days will he meet the other man ?

Answer—13. *M. vii*, 211.

42. The shadow of a gnomon 12 *añgulas* high is in one place 15 *añgulas.* The gnomon being moved 22 *añgulas* further its shadow is 18. The difference between the tips of the shadows is 25 and the difference between the lengths of the shadows is 3. Find the height of the light ?

Answer—100. *C.* 318 ; *Ar.* 16 ; *L.* 245.

43. The shadow of a gnomon 12 *añgulas* high being lessened by a third part of the hypotenuse became 14 *añgulas.* Tell, quickly, mathematician, that shadow ?

Answer—$22\frac{1}{2}$. *V.* 141.

---

* *L*=the Līlāvatī, *V*=Vīja Gaṇita, both by Bhāskara, *M*=Mahāvira, *S*=Srīdhara, *C*=Chaturveda.

44. Tell the perpendicular drawn from the intersection of strings mutually stretched from the roots to summits of two bamboos fifteen and ten *hastas* high standing upon ground of unknown extent ?

Answer—6. *L*. 160.

45. Of a quadrilateral figure whose base is the square of four and the face two *hastas* and altitude twelve, the flanks thirteen and fifteen, what is the area ?

Answer—108. *S*. 77.

46. In the figure of the form of a young moon the middle length is sixteen and the middle breadth is three *hastas*. By splitting it up into two triangles tell me, quickly, its area ?

Answer—24. *S*. 83.

47. The sides of a quadrilateral with unequal sides are 13×15, 13×20 and the top side is the cube of 5 and the bottom side is 300. What are all the values here beginning with that of the diagonals ?

Answers—315, 280, 48, 252, 132, 168, 224, 189, 44100.

M. *vii*, 59.

If $A^2 + B^2 = C^2$, and $a^2 + b^2 = c^2$ then the quadrilateral Ac, Bc, aC, bC is cyclic and the diagonals are $Ab + aB$ and $Aa + Bb$, the area is $\frac{1}{2}(ABc^2 + abC^2)$, &c. In the present case $A = 15$, $B = 20$, $C = 25$; $a = 5$, $b = 12$, $c = 13$. The diagonals are 315, 280 ; the area 44100. For full details see the *Līlāvatī*, § 193.

48. O friend, who knowest the secret of calculation, construct a derived figure with the aid of 3 and 5 as elements, and then think out and mention quickly the numbers measuring the perpendicular side, the other side and the hypotenuse ?

Answer—16, 30, 34. *M*. *vii*, 94.

That is construct a triangle of the form $2mn$, m^2-n^2, m^2+n^2, where $m = 5$, $n = 3$.

---

* L=the Līlāvatī, V=Vīja Gaṇita, both by Bhāskara, M=Mahāvira, S=Srīdhara, C=Chaturveda.

49. In the case of a longish quadrilateral figure the perpendicular side is 55, the base is 48 and then the diagonal is 73. What are the elements here ?

Answer—3, 8. *M. vii*, 121.

50. Intelligent friend, if thou knowest well the spotless Līlāvatī, say what is the area of a circle the diameter of which is measured by seven, and the surface of a globe or area like a net upon a ball, the diameter being seven, and the solid content within the same sphere ?

Answer—Area 38 $\frac{2423}{5000}$; surface 153 $\frac{1173}{1250}$; volume 179 $\frac{1487}{2500}$
L. 204.

51. In a circle whose diameter is ten, what is the circumference ? If thou knowest, calculate, and tell me also the area ?

Answer—$\sqrt{1{,}000}$, $\sqrt{6250}$. *S*. 85.

52. The measure of Rāhu is 52, that of the moon 25, the part devoured 7.

Answer—The arrow of Rāhu is 2, that of the moon 5.
C. 311.

This is an eclipse problem and means that circles of diameters 52 and 25 intersect so that the portion of the line joining the two centres common to the two circles is 7. The common chord cuts this into segments of 5 and 2.

53. The combined sum of the measure of the circumference, the diameter and the area is 1116. Tell me what the circumference is, what the calculated area, and what the diameter ?

Answer—108, 972, 36. *M. vii*, 32.

The rule given is circumference $=\sqrt{12\,(\text{combined sum}+64)}-\sqrt{64}$ which assumes that $\pi=3$.

---

* *L*=the Līlāvatī, *V*=Vīja Gaṇita, both by Bhāskara, *M*=Mahāvira, *S*=Srīdhara, *C*=Chaturveda.

54. The circumferential arrows are 18 in number. How many are the arrows in the quiver ?

Answer—37. *M*. 289.

The rule given is $n = \frac{(c+3)^2 + 3}{12}$ where c is the number of arrows in the outside layer.

55. Tell me, if thou knowest, the content of a spherical piece of stone whose diameter is a *hasta* and a half ?

Answer—$1\frac{25}{32}$. *S*. 93.

The rule given is $v = d^3 \left(\frac{1}{2} + \frac{1}{2.18}\right)$.

56. A sacrificial altar is built of bricks 6 *añgulas* high, half a *hasta* broad and one *hasta* long. It is 6 *hastas* long, 3 *hastas* broad and half a *hasta* high. Tell me rightly, wise man, what its volume is and how many bricks it contains.

Answer—9, 72. *S*. 96.

24 añgulas = 1 hasta.

57. If thou knowest, tell me quickly the measure of a mound of grain whose circumference is 36 and height 4 *hastas* ?

Answer—144. *Ś*. 102.

The rule used assumes that $\pi = 3$.

58. In the case of a figure having the outline of a bow, the string measure is 12, and the arrow measure is 6. The measure of the bow is not known. Find it, O friend.

Answer—$\sqrt{360}$. *M*. *vii*, 75.

59. In the case of a figure having the outline of a bow the string is 24 in measure, and its arrow is taken to be 4 in measure. What is the minutely accurate value of the area ?

Answer—$\sqrt{5760}$. *M*. *vii*, 72.

The rule used is $S = \frac{ac\sqrt{10}}{4}$.

* * * * *

* * * * *

* * * * *

* *L* = the Līlāvatī, *V* = Vīja Gaṇita, both by Bhāskara, *M* = Mahāvira, *S* = Srīdhara, *C* = Chaturveda.

60. Multiplier consisting of surds two, three and eight: multiplicand the surd three with the rational number five. Tell quickly the product ?

Answer—$\sqrt{9}+\sqrt{450}+\sqrt{75}+\sqrt{54}$. *V*. 32.

61. What is the number which multiplied by five and having the third part of the product subtracted, and the remainder divided by ten; and one-third, one half and a quarter of the original quantity added gives two less than seventy ?

Answer—48. *L.* 51.

The solution may be summarised this: $f(x)=68$, $f(3)=17/4$ therefore $x=\frac{3\times 68}{17/4}=48$.

62. The eighth part of a troop of monkeys squared was skipping in a grove and delighted with their sport. Twelve remaining were seen on the hill amused with chattering to each other. How many were there in all ?

Answer—48 or 16. *V*. 139.

63. The fifth part of the troop less three, squared, had gone to a cave and one monkey was in sight, having climbed on a branch. Say how many there were ?

Answer—50 or 5. V. 140.

"But," Bhāskara says, "the second is not to be taken for it is incongruous. People do not approve a negative absolute number."

64. Say quickly what the number is which added to five times itself divided by thirteen becomes thirty ?

Answer—$\frac{65}{3}$. *V*. 168.

65. A certain unknown quantity is divided by another. The quotient added to the divisor and the dividend is fifty-three. What is the divisor ?

Answer—5, 8. *M. vi*, 274.

---

* *L*=the Lilāvatī, *V*=Vīja Gaṇita, both by Bhāskara, *M*=Mahāvira, *S*=Srīdhara, *C*=Chaturveda.

66. What number is that which multiplied by nought and added to half itself and multiplied by three and divided by nought amounts to the given number sixty-three ?

Answer—14. This assumes that $\frac{0}{0}=1$. *L.* 46.

67. What four numbers are such that their product is equal to twenty times their sum, say, learned mathematician who art conversant with the topic of the product of unknown quantities ?

Answer—5, 4, 2, 11. *V.* 210.

Bhāskara puts arbitrary values for three of the quantities and gets 11 for the fourth.

68. If you are conversant with operations of algebra tell the number of which the fourth power less double the sum of the square and of two hundred times the simple number is ten thousand less one ?

Answer—11. *V.* 138.

This may be expressed by $x^4-2\ (x^2+200\ x)=9999$. It is the only case in which the fourth power occurs.

69. The square of the sum of two numbers added to the cube of their sum is equal to twice the sum of their cubes ?

Answer—1, 20 ; 5, 76, etc. *V.* 178.

70. Tell me, if you know, two numbers such that the sum of them multiplied respectively by four and three may when added to two be equal to the product of the same numbers ?

Answer—5, 10 and 11, 6. *V.* 209, 212.

71. Say quickly, mathematician, what is the multiplier by which two hundred and twenty-one being multiplied and sixty-five added to the product, the sum divided by a hundred and ninety-five becomes cleared ?

Answer—5, 20, 35 &c. *L.* 253 ; *V.* 65.

---

* *L*=the Līlāvatī, *V*=Vīja Gaṇita, both by **Bhāskara**, *M*=Mahāvira, *S*=Srīdhara, *C*=Chaturveda.

72. What number divided by six has a remainder of five, divided by five has a remainder of four, by four a remainder of three and by three one of two ?

Answer—59. *Br. xviii*, 7 ; *V.* 160.

73. What square multiplied by eight and having one added to the product will be a square ?

V. 82.

Here $8u^2+1=t^2$ and $u=6$, 35, etc. $t=17$, 99, etc.

74. Making the square of the residue of signs and minutes on Wednesday multiplied by ninety-two and eighty-three respectively with one added to the product an exact square ; who does this in a year is a mathematician.

Br. xviii, 67.

(1) $92\,u^2+1=t^2$ (2) $83\,u^2+1=t^2$.

Answer—(1) $u=120$, $t=1151$. (2) $u=9$, $t=82$.

75. What is the square which multiplied by sixty-seven and one being added to the product will yield a square-root ; and what is that which multiplied by sixty-one with one added to the product will do so likewise ? Declare it, friend, if the method of the 'rule of the square' be thoroughly spread, like a creeper, over thy mind ?

V. 87.

(1) $67\,u^2+1=t^2$. (2) $61\,u+1=t^2$.

Answers—(1) $u=5967$, $t=48842$. (2) $u=226,153,980$, $t=1,766,319,049$.

76. Tell me quickly, mathematician, two numbers such that the cube-root of half the sum of their product and the smaller number, and the square-root of the sum of their

---

* *L*=the Lilāvatī, *V*=Vīja Gaṇita, both by Bhāskara, *M*=Mahāvīra, =Sridhara, *C*=Chaṭurveda.

squares, and those extracted from the sum and difference increased by two, and that extracted from the difference of their squares added to eight, being all five added together may yield a square-root—excepting, however, six and eight?

V. 190.

$$\sqrt[3]{(xy+y)/2}+\sqrt{x^2+y^2}+\sqrt{x+y+2}+\sqrt{x-y+2}+\sqrt{x^2-y^2+8}=t^2$$

Answers—$x=8$; 1677/4 ; 15128, etc. ; $y=6$, 41 ; 246, etc.

---

* *L*=the Lîlâvatî, *V*=Vîja Gaṇita, both by Bhâskara, *M*=Mahâvira, *S*=Srîdhara, *C*=Chaṭurveda.

CHRONOLOGY.

		Circ.
Pythagoras	*B.C.*	530
Euclid	,,	290
Archimedes	,,	250
THE S'ULVASŪTRAS ..	,,	?
The Nine Sections ..	,,	150
Hipparchus	,,	130
Nicomachus	*A.D.*	100
Ptolemy	,,	150
Sassanian period begins ..	,,	229
The Sea Island Arithmetic ..	,,	250
Gupta period begins ..	,,	320
Diophantus	,,	360
Hypatia	,,	410
Bœthius	,,	470 b.
ĀRYABHATA	,,	476 b.
Damascius	,,	480 b.
Athenian schools closed ..	,,	530
Chang ch'iu-chien ..	,,	550
VARĀHA MIHIRA	,,	587 d.
BRAHMAGUPTA	,,	598 b.
Fall of Alexandria ..	,,	640
Gupta period ends ..	,,	650
Sassanian period ends ..	,,	652
Muhammad b. Mūsā ..	,,	820 d.
MAHĀVĪRA	,,	850 ?
'Tabit b. Qorra	,,	826 ?
el-Battānī	,,	877 b.
el-Birūni	,,	973 b.
Ibn Sīna	,,	980 b.
S'RĪDHARA	,,	991 b.
el-Karchi	,,	1016
'Omar b. Ibrāhīm el-Chaijāmi	,,	1046 b.
BHĀSKHARA	,,	1114 b.

BIBLIOGRAPHY.

(For a more complete bibliography see that given in the *Journal of the Asiatic Society of Bengal*, VII, 10, 1911.)

First Period.

Thibaut, G.—On the S'ulvasūtras, *J.A.S.B.*, XLIV, 1875.—The Baudhāyana S'ulvasūtra, *The Pandit* (Benares) 1875-6.—The Kātyāyana S'ulvasūtra, *Ib.*, 1882.

Bürk, A.—Das Apastamba-S'ulba-Sūtra, *Z.D.M.G.*, 55, 1901 ; 56, 1902.

Second Period.

Burgess, E. and Whitney, G.—The Sūrya Siddhānta, *Jour. Am. Or. Soc.*, VI, 1855.

Bapu Deva Sastri and Wilkinson, L.—*The Sūrya Siddhānta and the S'iddhānta S'iromani*, Calcutta, 1861.

Thibaut, G. and Sudharkar Dvivedi.—*The Pañchasiddhāntikā of Varaha Mihira*, Benares, 1889.

Rodet, L.—*Leçons de Calcul d'Aryabhaṭa*, Paris, 1879.

Kaye, G. R.—Āryabhaṭa, *J.A.S.B.*, IV, 17, 1908.

Third Period.

Colebrooke, H. T.—*Algebra with Arithmetic and Mensuration from the Sanscrit of Brahmagupta and Bhascara*, London, 1817.

Rangacarya, M.—*The Gaṇita-Sāra-Saṅgraha of Mahāvīracārya*, Madras, 1908.

RAMANUJACHARIA, N. AND KAYE, G. R.—The Triśatikā of S'rīdharāchārya, *Bib. Math.*, XIII, 3, 1913.

NOTATIONS.

BUHLER, G.—*Indische Palæographie*, Strassburg, 1896.

BAYLEY, E. C.—*On the Genealogy of Modern Numerals*, London, 1882.

WOEPCKE, F.—Mémoire sur la propagation des Chiffres indiens, *Jour. Asiatique*, 1863.

KAYE, G. R.—Indian Arithmetical Notations, *J. A. S. B.*, III, 7, 1907. The Use of the Abacus in Ancient India, *J. A. S. B.*, IV, 32, 1908. Old Indian Numerical Systems, *Indian Antiquary*, 1911.

FLEET, J. F.— Āryabhaṭa's system of expressing Numbers, *J. R. A. S.*, 1911. The Use of the Abacus in India, *J. R. A. S.*, 1911.

SMITH, D. E. AND KARPINSKI, L. C.—*The Hindu Arabic Numerals*, Boston, 1911.

OTHER WORKS.

SACHAU, E. C.—*Alberuni's India*, London, 1910.

THIBAUT, G.—Astronomie, Astrologie und Mathematik, *Grundriss der Indo-Arischen Philologie*, III, 9, 1899.

HOERNLE, R.—The Bakhshālī Manuscript, *Indian Antiquary*, XVIII, 1888.

KAYE, G. R.—Notes on Hindu Mathematical Methods, *Blb. Math.*, XI, 4, 1911.—Hindu Mathematical Methods, *Indian Education*—1910-1913. The Source of Hindu Mathematics, *J. R. A. S.*, 1910. The Bakhshālī Manuscript, *J. A. S. B.*, VII, 9, 1912.

HEATH, T. L.—*Diophantus of Alexandria*, Cambridge, 1910.

ROSEN, F.—*The Algebra of Mohammed ben Musa*, London, 1831.

SUTER, H.—*Die Mathematiker und Astronomen der Araber und Ihre Werke*, Leipzig, 1900.

YOSHIO MIKAMI.—*The development of Mathematics in China and Japan*, Leipzig, 1912.

The general works on the history of mathematics by CANTOR, GUNTHER, ZEUTHEN, TANNERY and v. BRAUNMUHL and the articles by WOEPCKE, RODET, VOGT, SUTER and WIEDEMANN should also be consulted.

XI——82.

INDEX.